T0135611

Bibliografische Information der Deutschen Nationalbibliothek

Die Deutsche Nationalbibliothek verzeichnet diese Publikation in der
Deutschen Nationalbibliografie; detaillierte bibliografische Daten sind
im Internet über http://dnb.d-nb.de abrufbar.

ISBN 978-3-8325-1485-3

Logos Verlag Berlin
Comeniushof, Gubener Str. 47,
10243 Berlin
Tel.: +49 030 42 85 10 90
Fax: +49 030 42 85 10 92
INTERNET: http://www.logos-verlag.de

Magnetization Dynamics of Rare-Earth Doped Magnetic Films

Dissertation zur Erlangung des
Doktorgrades der Naturwissenschaften (Dr. rer. nat.)
der Fakultät Physik der Universität Regensburg

vorgelegt von

Michael Binder

aus Manching

durchgeführt am
Institut für Experimentelle und Angewandte Physik
der Universität Regensburg
unter Anleitung von
Prof. Dr. C. H. Back

Dezember 2006

Promotionsgesuch eingereicht am 13. Dezember 2006

Tag der mündlichen Prüfung: 30. Januar 2007

Die Arbeit wurde angeleitet von Prof. Dr. C. H. Back

Prüfungsausschuss: Prof. Dr. G. Bali   (Vorsitzender)
                      Prof. Dr. C. H. Back   (1. Gutachter)
                      Prof. Dr. C. Schüller   (2. Gutachter)
                      Prof. Dr. S. Ganichev   (Prüfer)

Viele Probleme erscheinen uns nur deshalb so groß,
weil wir sie mit zu wenig Abstand betrachten.

Jochen Mariss

Contents

Glossary

Abbreviations

BLS	Brillouin Light Scattering
FFT	Fast Fourier Transformation
FMR	Ferromagnetic Resonance
FWHM	Full Width at Half Maximum
GMR	Giant Magneto Resistance
HWHM	Half Width at Half Maximum
LL	Landau-Lifshitz
LLG	Landau-Lifshitz-Gilbert
MOKE	Magneto Optic Kerr Effect
MRAM	Magnetic Random Access Memory
PEEM	Photo Emission Electron Microscopy
PIMM	Pulsed Inductive Microwave Magnetometer
RBS	Rutherford Backscattering
RE	Rare-Earth
RT	room temperature
SHG	Second Harmonic Generation
SRT	Spin Reorientation Transition
S/N	Signal to Noise Ratio
SQUID	Superconducting Quantum Interference Device
TEM	Transmission Electron Microscopy
TMR	Tunnel Magneto Resistance
TR-MOKE	Time Resolved Magneto Optic Kerr Effect
VNA-FMR	Vector Network Analyzer Ferromagnetic Resonance
VSM	Vibrating Sample Magnetometery
XMCD	X-ray Magnetic Circular Dichroism

Important Symbols

		SI unit
A	exchange stiffness constant	[J/m]
$\boldsymbol{B}$	magnetic induction	[T]
$\mathcal{B}_{\text{eff}}$	effective magnetic induction	[T]
E_{F}	Fermi energy	[J]
f	frequency	[1/s]
g, g_{eff}	g-factor, effective g-factor	[1]
γ	$= g|e|/2m_{\text{e}}$, gyromagnetic ratio	[1/Ts]
γ_{eff}	$= g_{\text{eff}}|e|/2m_{\text{e}}$, effective gyromagnetic ratio	[1/Ts]
$\boldsymbol{H}$	magnetic field	[A/m]
H_{c}	coercive field	[A/m]
$\boldsymbol{H}_{\text{eff}}, \mathcal{H}_{\text{eff}}$	effective magnetic field	[A/m]
$\boldsymbol{H}_{\text{dc}}$	applied magnetic field	[A/m]
H_{FMR}	ferromagnetic resonance field	[A/m]
$\boldsymbol{k}$	wave vector	[1/m]
$\boldsymbol{M}$	magnetization	[A/m]
M_{s}	saturation magnetization	[A/m]
$\boldsymbol{m} = \boldsymbol{M}/M_s$	unit magnetization vector	[A/m]
$\boldsymbol{r}$	position	[m]
t	time	[s]
T_{c}	Curie temperature	[K]
T_{L}	angular momentum compensation temperature	[K]
T_{M}	magnetization compensation temperature	[K]
$\alpha, \alpha_{\text{eff}}$	(effective) Gilbert damping parameter	[1]
ΔH	(HWHM) field linewidth	[A/m]
ω	angular frequency $= 2\pi f$	[1/s]
$\Delta\omega$	(HWHM) frequency linewidth	[1/s]
τ	magnetic decay time	[s]
τ_{b}	magnetic background decay time	[s]
τ_{r}	reflectivity decay time	[s]

Physical Constants

e	$= 1.60217733 \cdot 10^{-19}\,\mathrm{C}$	elementary charge
m_e	$= 1.6749286 \cdot 10^{-31}\,\mathrm{kg}$	electron mass
μ_0	$= 4\pi \cdot 10^{-7}\,\mathrm{Vs/Am}$	permeability
k_B	$= 1.380658 \cdot 10^{23}\,\mathrm{J/K}$	Boltzmann constant

All equations and constants of the present work are in SI units. Nevertheless, some numerical quantities are also transferred to CGS units as this is still a commonly used unit system in magnetism. The following conversions are used:

$$1\,\mathrm{Oe} = 1000/4\pi\,\mathrm{A/m},$$
$$1\,\mathrm{emu/cm}^3 = 1000\,\mathrm{A/m}.$$

For further help to convert the equations between the two unit systems, the reader is specially referred to [1].

1 Introduction and Motivation

Magnetism has been used in different ways over centuries and is of utmost importance in our time. The first simple applications were compass needles for navigation. Today complex magnetic sensors and actuators are irreplaceable, e. g. in modern anti-lock braking systems of cars. Beyond that, ferromagnetism plays a key role in our modern world as the main data storage devices, tapes and hard disks, are based on magnetic media. Therefore, especially in the computer industry, big efforts are made to improve existing technologies and develop new products from latest research results. An impressive example is that already 10 years after the discovery of the Giant Magneto Resistance (GMR) [2, 3] in 1988, first hard disks with GMR read/write-heads were commercially available accelerating the already exponential growth of the storage density. The Tunnel Magneto Resistance (TMR) [4] made it possible to increase the areal storage density of magnetic devices even further.

The new field of spintronics combines properties that are based on the spin of the electron, like ferromagnetism, with the traditional electronics, that is based on the charge of the electron [5,6]. An example for such a device is the Magnetic Random Access Memory (MRAM) in which the data is stored similar to a hard disk in two magnetic states of GMR/TMR elements and the writing and reading is done by current pulses [7]. The advantage is that this kind of memory is non-volatile. Besides the storage density, the speed of manipulating the data is extremely important. Pushing the speed into the GHz regime requires a detailed knowledge of dynamic magnetic processes as precessional phenomena dominate the magnetic switching on that time scale. Information about the precessional frequency f, which is a key parameter, is therefore essential. A further important parameter of magnetic materials at high frequencies is the damping parameter α. It describes the energy dissipation of the magnetic system and the relaxation of the moving magnetization to an equilibrium position. Both parameters, f and α, influence the switching speed. However, up to now a detailed knowledge of the interplay of the microscopic processes involved in magnetic damping is still missing or under debate due to their complexity. Therefore, from both, the fundamental point of view as well as from the technological point of view, it is desirable to obtain a deeper insight into the fundamental processes.

For applications it is important to control the precessional frequency and the damping parameter of technologically relevant materials, e. g. magnetic metals. Thus, a lot of research has been accomplished in the last years to understand and manipulate α by means of adding capping layers [8] or modulating the film roughness [9]. A very important approach is to utilize the spin-orbit (SO) interaction as this is known to play an important role in the damping for metallic ferromagnets [10] since SO-coupling is a channel for energy transfer between the spin- and lattice-system. A stronger SO-coupling leads to a higher damping. In most 3d-based ferromagnetic materials the orbital part of the total angular momentum $\boldsymbol{J} = \boldsymbol{L} + \boldsymbol{S}$, with the orbital angular momentum $\boldsymbol{L}$ and the spin angular momentum $\boldsymbol{S}$, is small and the SO-coupling weak. A natural step is to modify damping by doping materials with otherwise low intrinsic damping with materials possessing large SO-coupling or orbital angular momentum. One of the early works using this approach can be found in Reidy et al. [11]. They use rare-earth (RE) doping to modify the damping behavior of Permalloy (NiFe-alloy), which is known to exhibit low intrinsic damping [12]. RE-ions couple antiferromagnetically to 3d-metal ions and therefore RE-3d-metal compounds are ferrimagnets. Ferrimagnets posses a magnetic compensation point and an angular compensation point. The compensation points are depending on the sample composition and the temperature and can therefore be adjusted. In the vicinity of the compensation points f and α are tuneable. This works well for Gd as in this case L=0 and the additional damping via the SO-coupling is essentially zero [13].

In this thesis a detailed examination of the influence of rare-earth doping on magnetization dynamics of magnetic films is presented.
The thesis is organized as follows:

Chapter 2 gives an overview over the most relevant concepts of magnetism for the topic of this thesis. The equations of motion for the magnetization will be introduced and the essential mechanisms for the magnetic damping of metallic ferromagnets will be briefly summarized. Two simple models that are applied to describe magnetization dynamics in all-optical laser pump-probe experiments will be discussed at the end of the chapter.

Chapter 3 includes information about the main aspects of the sample preparation and properties of the magnetic samples.

Chapter 4 provides details about the experimental techniques and setups. The two mayor techniques for the magnetization dynamics experiments, namely Ferromagnetic Resonance (FMR) and Time Resolved Magneto Optic Kerr Effect (TR-MOKE) are presented. The equations which are required for the evaluation of the data in the experimental parts of the thesis will be introduced.

Chapter 5 deals with the well known ferromagnet NiFe. An introduction to the measurement procedures and the data evaluation is given. The results of the various experimental approaches will be discussed and compared.

Chapter 6 contains the result of measurements of various RE-doped NiFe samples. Gd, Dy and Ho are chosen for the RE-doping in the presented studies. The RE content is varied in a range from 0 to 16%. The essential doping mechanism is the SO-coupling. Different experiments allow to separate intrinsic and extrinsic contributions to the magnetic damping.

Chapter 7 summarizes the results of a special ferrimagnetic material, namely CoGd. Gd is chosen as it has nearly zero orbital moment and the magnetic damping channel via the SO-coupling is minimized. This allows one to investigate the magnetization dynamics in the compensation region. It will be shown that the precessional frequency f and the damping parameter α can be tuned around the compensation points.

Chapter 8 presents experimental data of a multilayered FeGd sample. The additional damping of Gd due to the SO-coupling is negligible as in the case of CoGd. The sample is of special interest for TR-MOKE measurements as it has a canted magnetic state which enables an easy triggering for all-optical pump-probe experiments. A further feature of FeGd samples is a possible Spin Reorientation Transition (SRT) which may be feasible for a fast switching of the magnetization.

Chapter 9 closes the thesis with a summary and an outlook.

Appendix A contains an overview of static magnetic properties of the NiFe samples that are investigated in chapters 7 and 8. The data are necessary for the evaluation of the dynamic measurements.

Appendix B shows the list of publications.

2 Theoretical Concepts

In this chapter an introduction into the theoretical aspects of magnetism relevant for the topic of this thesis will be given. First the contributions to the free energy of a ferromagnet will be explained. In a second step the equations of motion for the magnetization will be presented and damping mechanisms will be discussed. Then a theoretical mean-field description of ferrimagnetism will be introduced. Two simple models that are applied to describe magnetization dynamics in all-optical laser pump-probe experiments and the corresponding excitation scheme will be presented at the end of the chapter.

2.1 Ferromagnetism

A ferromagnet has a non-vanishing spontaneous magnetization even in the absence of an applied magnetic field below a certain temperature, the Curie temperature T_c. The origin of this spontaneous magnetization is the exchange interaction which is a consequence of quantum mechanical principles [14]. A fully quantum mechanical description is necessary to calculate fundamental magnetic properties of a ferromagnet (e. g. the exchange integral). However, once the ferromagnetic order is established and the energy landscape of a ferromagnet is known, it is sufficient – in many cases – to use a classical continuum approach to describe the behavior of the magnetic system.

The basic object in magnetism is the magnetic moment, which is coupled to an angular momentum [15]. The magnetic moment of an electron consists of an orbital part and a spin part. The orbital part is connected to the orbital angular momentum of the electron. The spin part arises from the intrinsic angular momentum, i. e. the spin of the electron [16]. The magnitude of the (macroscopic) magnetization $\boldsymbol{M}(\boldsymbol{r})$ is equal to the saturation magnetization M_s at every position $\boldsymbol{r}$ in the magnetic body [17]. Thus, it is sufficient to use the magnetic unit vector $\boldsymbol{m} = \boldsymbol{M}/M_s$ to describe the direction of $\boldsymbol{M}(\boldsymbol{r})$ at the position r. m will also be referred to as (reduced) magnetization in the following. All magnetic field and magnetization vectors of the formulas and equations to be presented depend on the position r and on the time t in the dynamic equations.

2.2 Magnetic Energies

The total free energy E_{tot} of the ferromagnetic materials investigated within this thesis is mostly determined by four energy contributions,

$$E_{tot} = E_z + E_{ex} + E_a + E_d, \tag{2.1}$$

with the Zeeman energy E_z, the exchange energy E_{ex}, the anisotropy energy E_a and the demagnetizing energy E_d. The corresponding energy density functional is defined as

$$\mathcal{E}_{tot} = \frac{E_{tot}}{V} = \frac{E_z}{V} + \frac{E_{ex}}{V} + \frac{E_a}{V} + \frac{E_d}{V} = \mathcal{E}_z + \mathcal{E}_{ex} + \mathcal{E}_a + \mathcal{E}_d, \tag{2.2}$$

where V is the volume of the sample.

It was shown by Brown [17] that the static equilibrium conditions for a magnetic system can be obtained from

$$\boldsymbol{M} \times \boldsymbol{H}_{eff} = 0 \tag{2.3}$$

$$\boldsymbol{H}_{eff} = -\frac{1}{\mu_0 M_s} \frac{\partial \mathcal{E}_{tot}}{\partial \boldsymbol{m}}, \tag{2.4}$$

with the total internal magnetic field $\boldsymbol{H}_{eff}$ and the permeability μ_0. $\boldsymbol{H}_{eff}$, which is acting on the magnetic moments inside a solid, can be obtained from the functional derivative of the total energy density $\mathcal{E}_{tot}$ with respect to the reduced magnetization $\boldsymbol{m}$. $\boldsymbol{M} \times \boldsymbol{H}_{eff}$ describes a torque that is exerted by the effective magnetic field on $\boldsymbol{M}$ [18]. When $\boldsymbol{M}$ and $\boldsymbol{H}_{eff}$ are collinear, no torque is exerted on $\boldsymbol{M}$. If the torque is zero and $\mathcal{E}_{tot}$ has a minimum, a possible equilibrium position for the magnetization is found. Note that $\boldsymbol{H}_{eff}$ depends on the magnetization and is consequently time dependent as well. It is convenient to use the magnetic field $\boldsymbol{H}$ instead of the magnetic induction $\boldsymbol{B}$, as, by virtue of $\boldsymbol{B} = \mu_0(\boldsymbol{H} + \boldsymbol{M})$, only $\boldsymbol{H}$ exerts a torque on $\boldsymbol{M}$.

The knowledge of the energy landscape is important to identify possible equilibrium states and to describe and understand the dynamic behavior and the known hysteretic effects of ferromagnetic system. In the following the energy terms will be shortly introduced. A nice introduction into micromagnetism can be found in [19].

Zeeman Energy

The interaction between an external magnetic field $\boldsymbol{H}$ and the magnetization $\boldsymbol{M}$ causes an energy contribution

$$E_{\mathrm{z}} = -\mu_0 M_{\mathrm{s}} \int_V \boldsymbol{m} \cdot \boldsymbol{H} \, dV. \tag{2.5}$$

An alignment of the magnetization along the direction of the field is favored as the Zeeman energy has a minimum in this case.

Exchange Energy

The existence of the exchange energy is a purely quantum mechanical phenomenon and a consequence of the Pauli exclusion principle and electrostatic interactions [15]. According to the Heisenberg model, the exchange energy is

$$E_{\mathrm{ex}} = -\sum_{ij} J_{ij} \, \boldsymbol{S}_i \, \boldsymbol{S}_j, \tag{2.6}$$

with spins $\boldsymbol{S}_i$, $\boldsymbol{S}_j$ and the exchange constant (or exchange integral) J_{ij} between the ith and j^{th} spin. $J_{ij} > 0$ favors parallel alignment of the spins, hence the coupling is ferromagnetic. For $J_{ij} < 0$ antiparallel alignment of the spins is preferred and the coupling is antiferromagnetic. J_{ij} is assumed to a be a constant for nearest neighbor atoms, and to be zero otherwise. Consequently, the exchange interaction is very short ranged. In solid state ferromagnets electrons can be localized or delocalized and thus the corresponding magnetic moments (spins) are also either localized (e. g. in 4f-metals) or delocalized (e. g. in 3d-metals). Thus different theoretical approaches are necessary to describe the systems adequately [20]. The Heisenberg model is simple but only valid for localized spins. In the 3d-ferromagnets (Ni, Fe, Co), which are used in this thesis, the role of the delocalized conduction electrons can not be neglected. The correct description needs to take into account both, the localized and the band character of the electrons (spins) [15, 21]. The length scale on which the exchange interaction is dominant is the so-called exchange length [18]

$$\Lambda = \sqrt{2A/\mu_0 M_{\mathrm{s}}^2}, \tag{2.7}$$

with the material specific exchange stiffness constant A, that is proportional to the exchange constant J_{ij}. The dimensions of the magnetic distributions for all samples in this thesis are larger than the exchange length. Thus a continuum approximation in which the discrete nature of the lattice is ignored and the microscopic origin of the exchange interaction is 'hidden' can be used [15, 22]. In this case expression (2.6) can be written as

$$E_{\text{ex}} = -A \int_V (\text{grad } \boldsymbol{m})^2 dV, \qquad (2.8)$$

Here the exchange stiffness constant A characterizes the interaction of the magnetic moments. Any deviation from the parallel (antiparallel) alignment of the spins enhances the exchange energy contribution in the case of ferromagnetic (antiferromagnetic) coupling.

Anisotropy Energy

The magnetization in most ferromagnetic materials has energetically preferred directions that are defined by the structure and symmetry of the crystal. The dependence of the magnetic energy on the orientation of the magnetization with respect to the crystallographic axes is referred to as magneto-crystalline anisotropy. The anisotropies reflect the symmetry of the crystal and originate in the spin-orbit coupling: the lattice potential leads to a certain alignment of the electron orbital and the spin that is coupled to the magnetic orbital moment of the electron is therefore also aligned along the corresponding direction. One distinguishes between cubic, orthorhombic and uniaxial anisotropy [19]. In polycrystalline ferromagnetic materials, e. g. NiFe, a uniaxial anisotropy can be induced during the sample deposition by an external applied magnetic field. This kind of anisotropy is often referred to as field induced anisotropy. Anisotropies lead to so-called easy and hard axes. The first corresponds to the lowest, the latter to the highest anisotropy energy. Hence, the magnetization tends to be aligned along an easy axis. The contributions due to magneto-crystalline anisotropies are not important in this thesis. The shape anisotropy is based on the geometrical shape of the magnetic body and is connected to the demagnetizing energy. This is explained more detailed in the next paragraph.

Demagnetizing Energy

Generally a ferromagnet with finite volume and magnetization $\boldsymbol{M}$ generates a magnetic field $\boldsymbol{H}_\mathrm{d}$. At the surface of the magnetic body the magnetic component that is perpendicular to the surface generates magnetic surfaces charges. These charges are the origin of $\boldsymbol{H}_\mathrm{d}$. The magnitude of the surface charge density is given by the divergence of $\boldsymbol{M}$. Using $\nabla \cdot \boldsymbol{B} = 0$ leads to the relation

$$\nabla \cdot \boldsymbol{H} = -\nabla \cdot \boldsymbol{M}, \qquad (2.9)$$

since $\boldsymbol{B} = \mu_0(\boldsymbol{H} + \boldsymbol{M})$. Equation (2.9) shows that the divergence of $\boldsymbol{M}$ gives rise to an opposite divergence of the magnetic field $\boldsymbol{H}_\mathrm{d}$. The field points opposite to the internal magnetic field of the sample. Hence, the field is named demagnetizing field. Outside the sample the field is often referred to as stray field. The demagnetizing energy can be written as

$$E_\mathrm{d} = \frac{1}{2}\mu_0 \int_{allspace} \boldsymbol{H}_\mathrm{d}^2 \, dV = -\frac{1}{2}\mu_0 M_\mathrm{s} \int_{sample} \boldsymbol{m} \cdot \boldsymbol{H}_\mathrm{d} \, dV, \qquad (2.10)$$

with the demagnetizing field $\boldsymbol{H}_\mathrm{d}$. The last term follows from the fact that $\boldsymbol{M} = 0$ outside the sample. Note that even though the last integral is only evaluated over the magnetic body, it includes the energies of the stray fields outside the magnetic body [15]. The first term reveals that the presence of a demagnetizing field increases the total energy. Consequently, the system tries to minimize stray fields. The demagnetizing fields can be very complicated functions of the position for an arbitrary shape of the ferromagnet. Nevertheless, $\boldsymbol{H}_\mathrm{d}$ is simple for an ellipsoidal ferromagnet [15, 23]. In this case it is uniform inside the magnet:

$$\boldsymbol{H}_\mathrm{d} = -\hat{N}\,\boldsymbol{M} \qquad (2.11)$$

with the demagnetizing tensor $\hat{N}$ that can be diagonalized if $\boldsymbol{M}$ is pointing along a principal axis of the ellipsoid:

$$\hat{N} = \begin{pmatrix} N_\mathrm{x} & 0 & 0 \\ 0 & N_\mathrm{y} & 0 \\ 0 & 0 & N_\mathrm{z} \end{pmatrix} \qquad (2.12)$$

where $\mathrm{Tr}\hat{N} = N_\mathrm{x} + N_\mathrm{y} + N_\mathrm{z} = 1$ is the trace of $\hat{N}$.

A thin magnetic film with infinite lateral extension in the x-y-plane (cp. Fig. 2.1) is a special case with $N_\mathrm{x} = N_\mathrm{y} = 0, N_\mathrm{z} = 1$.

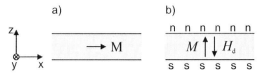

Figure 2.1: Cross section of a thin magnetic film with lateral dimensions much larger than the thickness and a homogenous magnetization M a) in the x-y-plane and b) perpendicular to the x-y-plane. In a) no demagnetizing field exists as the magnetic charges on the left and right surface are largely separated. In b) the magnetization generates magnetic poles on the upper and lower surface which lead to a demagnetizing field H_d oriented opposite to M.

As long as the magnetization lies in-plane, i. e. in the x-y-plane, no demagnetizing field occurs as the magnetic surface charges are only generated at the left and right edge which are infinitely separated and can be neglected (cp. Fig. 2.1 a)). In the perpendicular case (M parallel to the z-direction) the magnetic surface charges at the upper and lower surfaces lead to a demagnetizing field oriented opposite to the magnetization M (cp. Fig. 2.1 b)). These argumentations are valid for samples with lateral dimensions much larger than the thickness. This is the case for all samples presented in this thesis. The discussion reveals the shape anisotropy of thin magnetic films prefers an in-plane orientation of the magnetization. In this configuration the stray fields are minimized.

Note that the calculation of the demagnetizing field is the most time consuming problem in micromagnetic simulations.

All energy terms discussed above are important for the magnetic state of the ferromagnet. Usually magnetic domains are found in arbitrary ferromagnets below the Curie temperature T_c [21]. An externally applied magnetic field of certain magnitude is necessary to saturate the ferromagnet. The formation of domains is mainly driven by the minimization of the exchange energy and demagnetizing field energy. The short ranged exchange interaction (cp. eqn. (2.7)) tends to align neighboring spins parallel. On larger scales the stray field energy plays a more important role.

2.3 Magnetization Dynamics

Now the principles of magnetization dynamics will be introduced. The ferromagnetic ground state can be characterized by the energy landscape, i. e. by equations (2.3) and (2.4). The way to the lowest energy state is described by the appropriate equation of motion which governs the dynamics of the magnetization. Precessional dynamics of the magnetization take place on the timescale of nanoseconds and below. (Quasi)-static phenomena happen on the microsecond time range and above as the magnetization M is always in equilibrium with the effective field H_{eff} in this case. In the following the equation that governs the motion of M on the short timescales will be discussed.

2.3.1 The Landau-Lifshitz-Gilbert Equation

The time evolution of the magnetization in a ferromagnet was first addressed by L. Landau and E. Lifshitz in 1935 [22] in terms of the Landau-Lifshitz (LL) equation

$$\frac{dM}{dt} = -\gamma\mu_0(\,M \times H_{\text{eff}}) - \frac{\lambda}{M_{\text{s}}^2}\Big(\,M \times (\,M \times H_{\text{eff}})\Big), \qquad (2.13)$$

with the gyromagnetic ratio $\gamma = g|e|/\,2m_e$, where m_e is the mass of the electron, e the elementary charge and g = 2.0023, the g-factor for a free electron.

H_{eff} was already introduced in equation (2.4). Via H_{eff} all magnetic energies discussed enter the equation of motion. The first term on the right hand side describes the precessional motion of the magnetization due to the torque that is exerted by the effective field H_{eff}. Hence this term is often referred to as the precessional term. A precessional motion occurs as an angular momentum is inevitably coupled with the magnetic moment, i. e. with the magnetization M [18]. But the precessional term can not describe how the magnetization aligns with the effective field H_{eff}, i. e. how the magnetization reaches its equilibrium position. The second term describes the relaxation, i. e. the energy dissipation, of the magnetization, hence it is referred to as the damping term. The factor $\lambda = 1/\tau$ is a phenomenological damping constant which is proportional to the inverse relaxation time τ. In 1955, Gilbert modified the LL-equation [24, 25] to remove unphysical solutions for large damping parameters [26]. A viscous damping term was introduced in which damping is proportional to the magnetization velocity (friction like damping). The result is the Landau - Lifshitz - Gilbert equation

$$\frac{d\boldsymbol{M}}{dt} = \underbrace{-\gamma\mu_0(\boldsymbol{M} \times \boldsymbol{H}_{\text{eff}})}_{\text{precessional term}} + \underbrace{\frac{\alpha}{M_{\text{s}}}\left(\boldsymbol{M} \times \frac{d\boldsymbol{M}}{dt}\right)}_{\text{damping term}} \qquad (2.14)$$

with α being a dimensionless and phenomenological damping parameter. Another equivalent form of the LLG-equation is often used,

$$(1 + \alpha^2)\frac{d\boldsymbol{M}}{dt} = -\gamma\mu_0(\boldsymbol{M} \times \boldsymbol{H}_{\text{eff}}) - \frac{\alpha\gamma\mu_0}{M_{\text{s}}}\left(\boldsymbol{M} \times (\boldsymbol{M} \times \boldsymbol{H}_{\text{eff}})\right). \qquad (2.15)$$

The two equations can be converted into each other [17]. The choice between the two versions of the LLG-equation is often based on mathematical convenience. In the case of small damping ($\alpha \ll 1$) the damping torques of the LL-equation and the LLG-equation are equivalent ($\lambda = \alpha\,\mu_0\gamma M_s$). For large damping (large λ) the LL-equation predicts that the magnetic system will loose energy quickly and will rapidly reach its low energy state, whereas the LLG equation (for large α) predicts that the energy loss and approach of the low-energy state will become increasingly slower. However, for most magnetic systems there is no need to distinguish between the LL- and LLG-equation, because α is fairly small. These systems can be accurately modelled by using one of both equations [27]. Note, that the LLG-equation is also the starting point for micromagnetic simulations [19, 27]. The time evolution of the magnetization $\boldsymbol{M}$ according to the LLG-equation is depicted in Fig. 2.2. This is a macroscopic picture, the so-called single-spin or macro-spin model [18].

The magnitude of $\boldsymbol{M}$ and the cone angle between the effective field $\boldsymbol{H}_{\text{eff}}$ and the magnetization $\boldsymbol{M}$ is not changing during the motion if no damping is present (cp. Fig. 2.2 a)). In the case of damping (cp. Fig. 2.2 b)), the magnitude of $\boldsymbol{M}$ again is constant whereas the cone angle gets smaller as the damping term ($\boldsymbol{M} \times d\boldsymbol{M}/dt$) is perpendicular to $\boldsymbol{M}$ and $d\boldsymbol{M}/dt$. Hence, $\boldsymbol{M}$ is pushed towards $\boldsymbol{H}_{\text{eff}}$. As soon as the magnetization and the effective field are aligned parallel the motion stops. The precessional frequency ω is determined by

$$\omega = \gamma\mu_0\,\boldsymbol{H}_{\text{eff}}. \qquad (2.16)$$

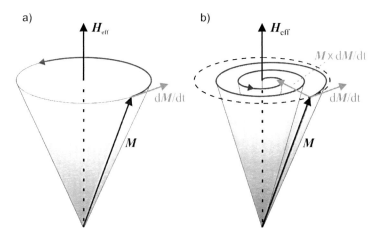

Figure 2.2: Illustration of the LLG equation: The magnetization M shall not be aligned parallel to the effective field H_{eff} in the beginning. a) Without damping, M precesses continuously around H_{eff} and describes the surface of a cone. b) With damping, M follows a helical trajectory towards H_{eff} until it is aligned parallel to H_{eff}.

Uniform Precession

In the following we want to examine the magnetization dynamics in the limit of the uniform precession (coherent precession) of the magnetization. This means that all magnetic moments (spins) precess with the same frequency and phase throughout the sample volume, i. e. contributions due to the exchange interaction are negligible. We focus on the special case of a thin magnetic film, considering only shape anisotropy ($N_x = N_y = 0, N_z \simeq 1$, cp. Fig. 2.1) and an applied magnetic field H_{dc}, i. e. the effective field $H_{\text{eff}} = H_{\text{dc}} - \hat{N} M$. Other contributions to H_{eff} like crystal anisotropies are neglected. For small motion angles of the magnetization the LLG-equation (2.14) can be linearized [28] and yields the precessional frequency ω for the uniform precession

$$\omega = \gamma \mu_0 \sqrt{H_x(H_x + N_z M_x) + (H_z - N_z M_z)^2}, \tag{2.17}$$

where H_x and H_z are the components of the applied magnetic field. H_y and M_y were chosen to be zero.

Equation (2.17) has the following limits

$$\omega = \gamma\mu_0(H_z - N_z M_s) \tag{2.18}$$

for $H_x = M_x = 0$ and $M_z = M_s$, i. e. the situation with the magnetization and applied field completely out-of-plane, and

$$\omega = \gamma\mu_0\sqrt{H_x(H_x + N_z M_s)} \tag{2.19}$$

for $H_z = M_z = 0$ and $M_x = M_s$, the in-plane equivalent.

Spin Waves

So far magnetization dynamics were discussed in the limit of the uniform precession. But generally the spin motion can also be non-uniform. Higher order excitations may exist in which the magnetic moments have the same frequency as in the uniform mode but different phases. A schematic drawing of a uniform precession and a spin wave is depicted in Fig. 2.3. This gives rise to various kinds of magnetic excitations referred to as spin waves or magneto-static waves. The concept of spin waves was introduced by Bloch [29] to explain the $T^{3/2}$-behavior of the saturation magnetization M_s of a three dimensional ferromagnet. As any proper wave, the spin waves are adequately described by a wave vector $\boldsymbol{k}$ with $|\boldsymbol{k}| = 2\pi/\lambda$, where λ is the wave length. The uniform precession mode, which is often also denoted as fundamental mode, is a special case of a spin wave, where the wave vector (and also the phase dispersion) is zero and the wave length is infinite. Spin waves can be classified by their wave length. Short wave length spin waves are exchange dominated and often referred to as 'spin waves'. This is fulfilled for wave vectors with $k \ll 1/\Lambda$, Λ being the exchange length (cp. eqn. (2.7)). Long wave length spin waves are dipole dominated and often simply referred to as 'magneto-static waves'.

The group velocity v_g and the phase velocity v_p are defined as [30]

$$v_g = \delta\omega/\delta k, \tag{2.20}$$
$$v_p = \omega/k. \tag{2.21}$$

If v_g and v_p have the same (opposite) signs the wave is called 'forward' ('backward') wave. Both, the propagation properties and the amplitude distributions of the spin waves depend on their propagation direction $\boldsymbol{k}$ with respect to the

Figure 2.3: Possible precessional modes in an one dimensional spin system: a) Fundamental mode, $\boldsymbol{k} = 0$ (spins precessing in phase) b) Higher order mode, propagating spin wave, $\boldsymbol{k} \neq 0$ (spins precessing out of phase with constant phase shift).

static magnetization $\boldsymbol{M}$ and the film plane [30,31].

Now we want to calculate the spin wave spectrum of a thin ferromagnetic sample. In the following θ_M denotes the angle of the static magnetization $\boldsymbol{M}$ with respect to the sample plane and θ_H denotes the angle of the applied magnetic field $\boldsymbol{H}$ with respect to the sample plane, respectively. If an out-of-plane field $\boldsymbol{H}$ is present the magnetization of an in-plane magnetized sample is dragging behind the external field due to the arising demagnetizing field. Neglecting anisotropies the angle of the magnetization θ_M can be calculated using the relation [32]

$$\sin(\theta_\mathrm{H} - \theta_\mathrm{M}) = \frac{M_\mathrm{s}}{2H} \sin(2\theta_\mathrm{M}). \tag{2.22}$$

Then the spin wave dispersion relation for a thin film is given by [32]

$$\begin{aligned}
\omega^2 = \omega_\mathrm{uni}^2 \; &- \; \frac{1}{2}\gamma^2\mu_0^2 M_\mathrm{s} k_\| d \big\{ [\cos^2\theta_\mathrm{M} - \sin^2\theta_\mathrm{M}\cos^2\theta_k][H\cos(\theta_\mathrm{H} - \theta_\mathrm{M}) \\
&- \; M_\mathrm{s}\sin^2\theta_\mathrm{M}] - \sin^2\theta_k[H\cos(\theta_\mathrm{H} - \theta_\mathrm{M}) + M_\mathrm{s}\cos(2\theta_\mathrm{M})]\big\} \\
&+ \; \gamma^2\mu_0 D k_\|^2[2H\cos(\theta_\mathrm{H} - \theta_\mathrm{M}) + M_\mathrm{s}(1 - 3\sin^2\theta_\mathrm{M})], \tag{2.23}
\end{aligned}$$

where $D = 2A/Ms$ is the spin wave stiffness constant and θ_k is the angle between the in-plane projection of the static magnetization and the propagation direction of the spin wave with wave vector $k_\|$ in the sample plane. ω_uni is the resonance frequency of the uniform precession of the magnetization which is given by [32]

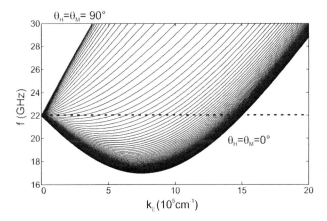

Figure 2.4: Calculated spin wave dispersion for a 30 nm thick NiFe film with $\theta_k = 0$ and $0° \leq \theta_H \leq 90°$ at a frequency of 22 GHz. The various curves correspond to a magnetic field incline by 1° steps. The intersection with the dashed line shows the wave vectors $k_{\parallel}^*(\theta_k)$ for magnon states that are degenerate with the uniform mode.

$$\omega_{\mathrm{uni}}^2 = \gamma^2 \mu_0^2 [H_0 \cos(\theta_H - \theta_M) - M_s \sin^2 \theta_M][H_0 \cos(\theta_H - \theta_M) + M_s \cos(2\theta_M)]. \quad (2.24)$$

In Fig. 2.4 the spin wave dispersion for a 30 nm thick NiFe sample for $\theta_k = 0$, $0° \leq \theta_H \leq 90°$ and a resonance frequency of 22 GHz is depicted.

It is obvious that a spin wave with a finite wave vector $k_{\parallel}^*(\theta_k)$ exists, which has the frequency of the uniform mode. This means that the uniform mode is degenerate with a $k_{\parallel}^*(\theta_k)$ mode. A detailed analysis of equation (2.23) reveals, that for $\theta_H = 90°$ no spin wave mode is degenerate with the uniform mode [32], irrespective of the the values of H, θ_H, θ_k and M_s. This will be important for two-magnon scattering, that is presented in the following subsection, as the uniform mode with the frequency f and $k_{\parallel} = 0$ can scatter into the degenerate spin wave mode with the same frequency f but $k_{\parallel}^*(\theta_k) \neq 0$.

The calculation of the spin wave spectrum using equation (2.23) does not account for Perpendicular Standing Spin Waves (PSSW) [32]. Another approach to calculate the spectrum of spin waves including the exchange interaction in an infinite ferromagnetic medium is the Herring-Kittel formula [33],

$$\omega = \gamma\mu_0 \left[\left(H + \frac{2A}{\mu_0 M_{\mathrm s}} k^2 \right) \left(H + \frac{2A}{\mu_0 M_{\mathrm s}} k^2 + M_{\mathrm s} \sin^2 \theta_k \right) \right]^{1/2}, \qquad (2.25)$$

with the three dimensional wave vector $\boldsymbol{k}$ and θ_k the angle between $\boldsymbol{k}$ and the magnetization $\boldsymbol{M}$. In a thin magnetic film with finite thickness d the spin wave spectrum is modified due to the fact that the translational invariance of an infinite medium is broken in the vicinity of the film surfaces [31, 34]. For an in-plane magnetized film with $\boldsymbol{k}_{\|} \perp \boldsymbol{M}$ and in the limit $k_{\|} d \ll 1$ the relation for the PSSW, also referred to as thickness or exchange dominated spin waves, is

$$\omega = \gamma\mu_0 \Bigg\{ \left(H + \frac{2A}{\mu_0 M_s} k_{\|}^2 + \frac{2A}{\mu_0 M_s} \left(\frac{p\pi}{d} \right)^2 \right) \cdot$$
$$\left(H + \left[\frac{2A}{\mu_0 M_s} + H \left(\frac{M_s/H}{p\pi/d} \right)^2 \right] k_{\|}^2 + \frac{2A}{\mu_0 M_s} \left(\frac{p\pi}{d} \right)^2 + M_s \right) \Bigg\}^{1/2},$$
$$k^2 = k_{\|}^2 + k_{\perp}^2 = k_{\|}^2 + \left(\frac{p\pi}{d} \right)^2, \quad p \geq 1, \qquad (2.26)$$

with the out-of-plane wave vector $\boldsymbol{k}_{\perp}$ (i.e. $\boldsymbol{k}_{\perp}$ is parallel to the sample surface normal) [31]. From equation (2.26) it is obvious that for the PSSW the wave vector is quantized in the direction perpendicular to the film surface, i. e. $k_{\perp} = p\pi/d$, $p \geq 1$ due to the limited thickness of the sample. Also the rather weak $f(k_{\|})$ dependence for $k_{\|} \ll p\pi/d$ can easily be seen. In Fig. 2.5 PSSW modes for fixed boundary conditions are sketched. More details about spin waves in magnetic samples are available in [30, 31, 35, 36].

2.3.2 Energy Dissipation - Damping

The damping parameter α has been introduced as a 'phenomenological' damping parameter, which implies that a detailed knowledge of the interplay of the microscopic processes involved in magnetic damping is still missing or under debate due to their complexity. This is one of the reasons why magnetization dynamics is a field of active research. In the following the most important mechanisms that contribute to magnetic damping in metallic ferromagnets will be discussed.

Magnetic damping can be separated into two contributions - intrinsic and extrinsic relaxation. Magnetic relaxation can be sample dependent and therefore not

a)

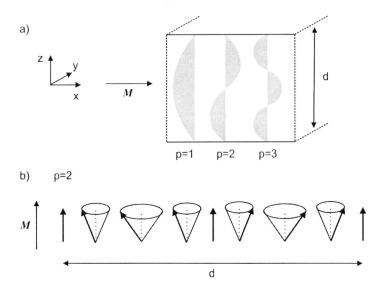

Figure 2.5: a) Scheme of PSSW modes for fixed boundary conditions in a thin film with thickness d [37]. The static magnetization M lies in-plane and is oriented along the x-axis. In b) the precessional motion of the spins around the static magnetization for the PSSW with p=2 is sketched.

necessarily reflects the intrinsic properties of the materials. Intrinsic damping includes all damping effects that are an integral part of the magnetic system, they can not be avoided, for example electron scattering with phonons and thermally excited magnons at finite temperatures [38,39]. Extrinsic damping contains all contributions that can be avoided, for example contributions of structural defects of the sample, complex geometries or non-uniform excitation. The smallest measured damping under well defined conditions is commonly considered to be the intrinsic damping of the system.

Intrinsic Damping
Different relaxation processes can lead to magnetic damping. For metallic ferromagnets the intrinsic damping is mostly caused by itinerant electrons and the spin orbit coupling [39,40]. There are two ways intrinsic damping in metals can

be understood.

First, Kamberský developed a model including the spin-orbit coupling based on the observation that a change in the direction of the magnetization changes the Fermi surface [10]. This can be described in a classical picture. The magnetization in ferromagnetic metals mostly originates from the spin angular momentum S, and the SO-coupling is proportional to $L \cdot S$ [15]. As the precession of the magnetization evolves in time and space, i. e. S changes, also the Fermi surface is distorted periodically in time and space, i. e. the orbital angular momentum L changes via the SO-coupling. This is often referred to as 'Breathing Fermi Surface'. The effort of the electrons to repopulate the changing Fermi surface is delayed by the finite relaxation time τ_{relax} of the electrons. This results in a phase lag between the precessing magnetization and the changing Fermi surface. This phase lag causes magnetic damping. More details about this model can be found in [39].

Second, the s-d-exchange interaction is important to understand the damping via itinerant electrons. The s-d-interaction can be viewed as two precessing magnetic moments corresponding to the localized d electrons and the itinerant electrons, which are mutually coupled by the s-d-exchange field. In the absence of damping the low energy excitation (acoustic or ferromagnetic resonance mode) corresponds to a parallel alignment of all magnetic moments precessing together in phase. Due to a finite spin mean free path of the itinerant electrons, spin relaxation via spin flips has to be considered. This results in a phase lag between the two precessing magnetic moments (s and d moments) [41], and consequently in magnetic damping. Again the reader is referred to [39] for further details.

The phase lag for the breathing Fermi surface and the s-d-interaction is proportional to the ferromagnetic resonance (FMR) frequency (cp. subsection 4.1.1). In both cases one obtains a friction like damping as described by the Gilbert relaxation term (cp. subsection 2.3.1). Detailed calculations are often complex and not easy to penetrate [39]. Kamberský showed that the intrinsic damping in ferromagnetic metals is treated more generally in the 'Breathing Fermi Surface' model [42]. Quantitative calculations [42–46] reveal, that the spin-orbit interaction indeed is the leading intrinsic magnetic damping mechanism for ferromagnetic metals. The quantitative calculations yield reasonable results for

NiFe, but for pure materials like Co, Fe and Ni the deviations to the experimental data are quite large.

Another contribution to intrinsic magnetic damping in metals is given by eddy currents which are caused by the precessing magnetization and /or rf-fields in ferromagnetic resonance experiments (cp. subsection 4.1.1). The damping can be neglected for thin films and reasonably low frequencies, the role of eddy currents is insignificant for NiFe samples with thicknesses below 100 nm [39]. Hence, a contribution is only expected for the thicker CoGd samples in this thesis.

Direct magnon-phonon scattering is another damping mechanism in ferromagnetic metals [47]. But the contribution to the damping in metallic ferromagnets (Fe, Co, Ni, NiFe) is very small (at least 30 times smaller than the intrinsic damping values of the materials) and can be neglected [40].

So far damping mechanisms that contribute to intrinsic damping were listed. In the following contributions to extrinsic damping will be introduced.

Extrinsic Damping

Extrinsic damping can be caused by defects and inhomogeneities of the structure and the magnetization of the sample. In the model of extrinsic damping by two-magnon scattering a uniform precession magnon $k = 0$ scatters into a magnon with $k \neq 0$. Due to energy conservation the uniform mode can only scatter into spin waves with the same precessional frequency, i. e. $\omega(0) = \omega(k)$ with k determined by the magnon dispersion relation. Momentum conservation has not to be fulfilled due to the loss of translational invariance in thin magnetic films. Two magnon scattering pumps the magnetic energy into other spin wave modes. Hence the magnetic excitation does not disappear as the energy stays in the magnetic system. In order to reach equilibrium the magnetic energy has to be transferred into the lattice by intrinsic damping [39]. The concept of two magnon scattering was successfully used to describe the behavior of different magnetic materials [48–50], and also of (ultrathin) metallic films [32, 51, 52].

The contribution of two-magnon processes to the magnetic damping can be influenced by the measurement geometry as the spin wave manifold is strongly dependent on the out-of-plane angle of the magnetization θ_M (cp. Fig. 2.4). For FMR measurements in the perpendicular configuration $\theta_H = \theta_M = 90°$ (cp. subsection 4.1.1), two-magnon scattering vanishes [32, 40]. Hence FMR measurements in the in-plane geometry ($\theta_{\mathrm{H}} = \theta_{\mathrm{M}} = 0°$) and the perpendicular geometry can be used to separate intrinsic damping components and extrinsic damping components that are caused by two-magnon scattering.

Another possible contribution to extrinsic damping is radial damping. This occurs in experiments in which the magnetic excitation is strongly localized like in time resolved magneto optic Kerr effect (TR-MOKE) experiments (cp. subsection 4.3.2). Higher order spin waves with finite wave length are generated and emitted to undisturbed regions of the sample. These waves carry energy and angular momentum away which leads to an increased damping in the excitation region. In addition, these spin waves can gain energy by two magnon processes. Therefore in optical pump-probe experiments a higher damping parameter as the intrinsic one may be found. In all-optical experiments Józsa [53] found an increase of the damping parameter α of about 25% compared to dynamic measurements using other techniques for 10 nm NiFe. The strength of the effect strongly depends on the propagation speed of the excited spin waves (cp. eqn. (2.20)) and thus on the spin wave dispersion relation.

An interesting extrinsic damping mechanism is the so-called spin-pumping effect [8]. The pumping of a spin current from a ferromagnetic layer away into a non-magnetic adjacent layer can result in an increased damping of the magnetization for very thin ferromagnets. This effect is negligible for this thesis as the thicknesses of the magnetic layers of our samples are at least 30 nm. Further details can be found in [39].

In conclusion one should keep in mind that the measured damping parameter α may consist of intrinsic and extrinsic components.

2.4 Ferrimagnetism

So far only ferromagnetism has been treated in this thesis. In the following section the concepts of ferrimagnetism will be introduced that are of special relevance for the understanding of the results on the CoGd-alloys in chapter 7. A ferrimagnet consists of two (or more) materials. In the simplest case the ferrimagnet is homogenous and contains two materials with a mixing ratio of 1:1. In this case the ferrimagnet can be described by two interpenetrating sublattices. The exchange constant is negative, $J_{ij} < 0$, which causes the nearest neighbor magnetic moments to lie antiparallel with respect to one another (cp. exchange energy in section 2.2). Thus the coupling between the sublattices is antiferromagnetic which is schematically depicted in Fig. 2.6. One sublattice consists of magnetic moments pointing in positive, the other sublattice of magnetic moments pointing in negative x-direction. The net-magnetic moment of the sample is the sum of the magnetic moments of the two sublattices. When the moments are equal for both lattices one gets an antiferromagnet and the magnetic moments cancel out each other. The net-magnetization of the alloy depends on the temperature and the composition of the sample.

In our experiments we are mainly interested in alloys of rare-earth and transition metals (TM). Here we will focus on the case of RE=Gd and TM=Co. In Fig. 2.7 the temperature dependence of the magnetization of the two sublattices and the net-magnetization for a CoGd-alloy are schematically shown.

At low temperatures the magnetization of the Gd is dominant and the net-magnetization points 'upwards'. With increasing temperature the magnetization of both sublattices gets smaller and so does the net-magnetization. At a certain temperature T_M, the magnetization compensation temperature, the net-magnetization vanishes. If the temperature is increased further the net-magnetization increases again and points 'downwards' as the magnetization of

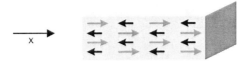

Figure 2.6: Illustration of a ferrimagnet consisting of two interpenetrating antiferromagnetically coupled sublattices with magnetic moments of different magnitude.

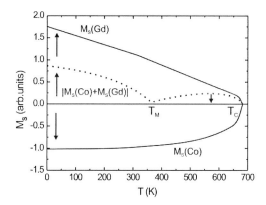

Figure 2.7: Scheme of the temperature dependence of a CoGd-alloy. The Co and Gd form two antiferromagnetically coupled sublattices. As the magnetizations of both are unequally depending on T, a magnetic compensation temperature T_M exists at which the net-magnetization is zero. At the Curie temperature T_c the magnetization finally vanishes.

the Co-sublattice exceeds the one of Gd. At the Curie temperature T_c the net-magnetization finally vanishes as well as the strongly exchange coupled magnetizations of the sublattices. Note that if strong coupling is assumed, like in the case of CoGd, there is just one Curie temperature for the whole magnetic system. This behavior has already been experimentally confirmed for a $Co_{0.779}Gd_{0.221}$ alloy [54,55]. The results are plotted in Fig. 2.8. For this composition the magnetization compensation temperature T_M is close to room temperature (RT). However, T_M can be tuned by varying the composition of the alloy.

In the following an equation of motion shall be deduced for a ferrimagnet with strong antiferromagnetic coupling between the magnetic moments. Starting with the LLG-equation (2.14) we formulate the LLG-equation for each sublattice. According to Weiss's model of an antiferromagnet we assume that the molecular field on one sublattice is proportional to the magnetization of the other sublattice [15,56]. Then the equations of motions for both sublattice magnetizations are

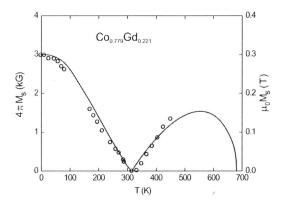

Figure 2.8: Temperature dependence of the saturation magnetization M_s of amorphous $Co_{0.779}Gd_{0.221}$. The open circles mark experimental results, the solid line was calculated via mean field theory [54, 55].

$$\frac{d\boldsymbol{M}_1}{dt} = -\gamma_1\mu_0\left(\boldsymbol{M}_1 \times \left[\boldsymbol{H}_\mathrm{eff} + \lambda\,\boldsymbol{M}_2\right]\right) + \frac{\alpha_1}{M_1}\left(\boldsymbol{M}_1 \times \frac{d\boldsymbol{M}_1}{dt}\right), (2.27)$$

$$\frac{d\boldsymbol{M}_2}{dt} = -\gamma_2\mu_0\left(\boldsymbol{M}_2 \times \left[\boldsymbol{H}_\mathrm{eff} + \lambda\,\boldsymbol{M}_1\right]\right) + \frac{\alpha_2}{M_2}\left(\boldsymbol{M}_2 \times \frac{d\boldsymbol{M}_2}{dt}\right), (2.28)$$

with $\boldsymbol{M}_1 = M_1\,\boldsymbol{m}_1$, $\boldsymbol{M}_2 = M_2\,\boldsymbol{m}_2$ and $\boldsymbol{m}_1, \boldsymbol{m}_2$ magnetic unit vectors. λ is the molecular field constant which is negative due to the antiferromagnetic coupling. It is assumed that $\boldsymbol{M}_1$ and $\boldsymbol{M}_2$ are always collinear and antiparallel (ferromagnetic resonance case), hence

$$\boldsymbol{M}_1 \times \boldsymbol{M}_2 = 0,$$
$$\boldsymbol{M}_1 = M_1\,\boldsymbol{m}_1,$$
$$\boldsymbol{M}_2 = M_2\,\boldsymbol{m}_2 = -M_2\,\boldsymbol{m}_1.$$

Defining $\boldsymbol{\mu}\,\|\,\boldsymbol{M}_1$ with $\boldsymbol{\mu}^2 = 1$ it follows:

$$\frac{M_1}{\gamma_1\mu_0}\frac{d\boldsymbol{\mu}}{dt} = -M_1(\boldsymbol{\mu} \times \boldsymbol{H}_\mathrm{eff}) + \frac{\alpha_1 M_1}{\gamma_1\mu_0}\left(\boldsymbol{\mu} \times \frac{d\boldsymbol{\mu}}{dt}\right),$$

$$-\frac{M_2}{\gamma_2\mu_0}\frac{d\boldsymbol{\mu}}{dt} = M_2(\boldsymbol{\mu} \times \boldsymbol{H}_\mathrm{eff}) + \frac{\alpha_2 M_2}{\gamma_2\mu_0}\left(\boldsymbol{\mu} \times \frac{d\boldsymbol{\mu}}{dt}\right).$$

Adding both equations yields:

$$\frac{1}{\mu_0}\left(\frac{M_1}{\gamma_1} - \frac{M_2}{\gamma_2}\right)\frac{d\boldsymbol{\mu}}{dt} = -(M_1 - M_2)(\boldsymbol{\mu} \times \boldsymbol{H}_{\text{eff}}) + \frac{1}{\mu_0}\left(\frac{\alpha_1 M_1}{\gamma_1} + \frac{\alpha_2 M_2}{\gamma_2}\right)\left(\boldsymbol{\mu} \times \frac{d\boldsymbol{\mu}}{dt}\right)$$

$$\frac{d\boldsymbol{\mu}}{dt} = \frac{-\mu_0(M_1 - M_2)}{M_1/\gamma_1 - M_2/\gamma_2}(\boldsymbol{\mu} \times \boldsymbol{H}_{\text{eff}}) + \frac{\alpha_1 M_1/\gamma_1 + \alpha_2 M_2/\gamma_2}{M_1/\gamma_1 - M_2/\gamma_2}\left(\boldsymbol{\mu} \times \frac{d\boldsymbol{\mu}}{dt}\right)$$

Defining the net magnetization vector $\boldsymbol{M} = M\boldsymbol{\mu}$ with $M = M_1 - M_2$ for the ferrimagnetic system one gets

$$\frac{d\boldsymbol{M}}{dt} = \frac{-\mu_0(M_1 - M_2)}{M_1/\gamma_1 - M_2/\gamma_2}(\boldsymbol{M} \times \boldsymbol{H}_{\text{eff}}) + \frac{\alpha_1 M_1/\gamma_1 + \alpha_2 M_2/\gamma_2}{M_1/\gamma_1 - M_2/\gamma_2}\frac{1}{M}\left(\boldsymbol{M} \times \frac{d\boldsymbol{M}}{dt}\right).$$

Finally defining α_{eff} and γ_{eff} this yields the LLG-equation for the strongly antiferromagnetically coupled ferrimagnet [57]

$$\frac{d\boldsymbol{M}}{dt} = -\gamma_{\text{eff}}\,\mu_0(\boldsymbol{M} \times \boldsymbol{H}_{\text{eff}}) + \frac{\alpha_{\text{eff}}}{M}\left(\boldsymbol{M} \times \frac{d\boldsymbol{M}}{dt}\right) \quad (2.29)$$

$$\text{with} \quad \gamma_{\text{eff}} = \frac{M_1 - M_2}{M_1/\gamma_1 - M_2/\gamma_2}\ , \quad \alpha_{\text{eff}} = \frac{\alpha_1 M_1/\gamma_1 + \alpha_2 M_2/\gamma_2}{M_1/\gamma_1 - M_2/\gamma_2}.$$

Thus, the ferromagnetic resonance (FMR) frequency is

$$\omega = |\gamma_{\text{eff}}|\mu_0 H_{\text{eff}}. \quad (2.30)$$

Extensive calculations that do not assume a collinear and antiparallel alignment of $\boldsymbol{M}_1$ and $\boldsymbol{M}_2$ were performed by Kaplan et al. [58] to determine the exchange resonance frequency ω_{ex} for the ferrimagnetic system. The exchange resonance frequency is given by

$$\omega_{\text{ex}} = |\mu_0\lambda(\gamma_1 M_2 - \gamma_2 M_1)|$$
$$= |\mu_0\lambda\gamma_1\gamma_2(M_1/\gamma_1 - M_2/\gamma_2)|. \quad (2.31)$$

Both resonance modes are schematically depicted in Fig. 2.9 for the case of an applied magnetic field H_{a} which is much smaller than the exchange fields $H_{\text{ex},i} = \lambda M_i$ $(i = 1, 2)$ (other contributions to $\boldsymbol{H}_{\text{eff}}$ and damping are neglected for simplicity).

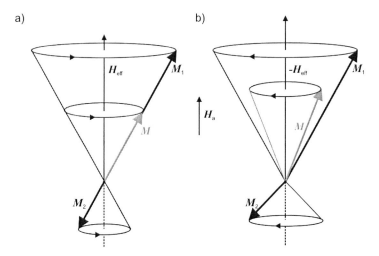

a)   b)

Figure 2.9: Schematic illustration of a) the low frequency ferromagnetic resonance and b) the high frequency exchange mode of a ferrimagnet for $H_a \ll H_{ex,1}, H_{ex,2}$ (other contributions to $\boldsymbol{H}_{eff}$ and damping are neglected). The sense of precession of both modes is opposite. The magnitude of $\boldsymbol{H}_{eff}$ in b) is much bigger in reality than in a) due to the large exchange fields $\boldsymbol{H}_{ex,1}$, $\boldsymbol{H}_{ex,2}$. This causes the high frequency of the exchange mode. Note that in b) - $\boldsymbol{H}_{eff}$ is plotted.

In the ferromagnetic resonance mode (cp. Fig. 2.9 a)) the applied field $\boldsymbol{H}_a$ governs the behavior of the system. $\boldsymbol{M}_1$ and $\boldsymbol{M}_2$ are exactly antiparallel. The torque of the bigger magnetization $\boldsymbol{M}_1$ determines the direction of the precession of the net-magnetization $\boldsymbol{M}$. The exchange fields $\boldsymbol{H}_{ex,1}$ and $\boldsymbol{H}_{ex,2}$ that are antiparallel to $\boldsymbol{M}_1$ and $\boldsymbol{M}_2$ (cp. eqns. (2.27) and (2.28)), respectively, do not influence the dynamics of the total system as these fields are always (anti-)parallel to the net-magnetization $\boldsymbol{M}$. In the higher frequency exchange mode $\boldsymbol{M}_1$ and $\boldsymbol{M}_2$ are not antiparallel anymore and consequently $\boldsymbol{M}$ is not (anti-)parallel to $\boldsymbol{M}_1$ and $\boldsymbol{M}_2$. This mode is governed by the exchange fields. As the exchange field $\boldsymbol{H}_{ex,1}$ governs the effective field $\boldsymbol{H}_{eff}$, the direction of $\boldsymbol{H}_{eff}$ and therefore of the direction of the precession of the net magnetization $\boldsymbol{M}$ changes. Usually the frequency of the exchange mode is much higher than the frequency of the ferromagnetic resonance mode as $H_a \ll H_{ex}$.

It is important to keep in mind that the LLG-equation for ferrimagnets (2.29), the according ferromagnetic precession frequency (cp. eqn. (2.30)) and the exchange frequency (cp. eqn. (2.31)) depend on the temperature and the composition of the sample. This is due to the fact that they contain the magnetizations M_1 and M_2 of the two sublattices which are temperature and composition dependent.

A closer look at α_{eff} and γ_{eff} (cp. eqn. (2.29)) reveals two interesting aspects. Both denominators are equal and have a root. At a fixed composition the temperature at which the denominator is zero, i. e. $(M_1/\gamma_1) = (M_2/\gamma_2)$, is called the angular compensation temperature T_{L} [1]. The angular momentum vanishes at the angular momentum compensation point.

At T_{L} the damping of the system α_{eff} and the gyromagnetic ratio γ_{eff} of the ferromagnetic resonance mode are expected to diverge. This is illustrated in Fig. 2.10 a) for one possible case. The corresponding ferromagnetic resonance frequency ω according to equation (2.30) and the exchange mode frequency ω_{ex} according to equation (2.31) are schematically depicted in Fig. 2.10 b). A more detailed analysis of equation (2.29) shows that the sense of the precession of M is constant for $T < T_{\text{L}}$. At T_{L} the sense of the precession changes sign as $(M_1/\gamma_1) - (M_2/\gamma_2)$ changes sign and is again constant for $T > T_{\text{L}}$. The sense of the precession does not change at T_{M} as the negative sign of γ_{eff} is compensated by the inverted direction of M in equation (2.29) for $T_{\text{M}} < T < T_{\text{L}}$. The negative value of α_{eff} for $T > T_{\text{L}}$ is also correlated to the inversion of the sense of the precession at T_{L}.

The phenomenological mean-field damping parameter α_{eff} governs how quickly the system as a whole dissipates energy. Hence, an ideal ferrimagnet should dissipate angular momentum instantaneously at T_{L} [57]. At the same time the frequency of the exchange mode is expected to tend to zero, as equation (2.31) contains a term that is equal to the denominators of α_{eff} and γ_{eff}.

It is also possible that the magnetization of both sublattices cancel out each other and therefore the numerator of γ_{eff} is zero, i. e. $M_1 = -M_2$. This happens at the magnetization compensation temperature T_{M} (cp. Fig. 2.10). At this point the FMR frequency should be zero. At the same time the coercive field

---

[1]In atoms the orbital magnetic moment μ associated with an orbiting electron is connected to the angular momentum L of the electron via $\mu = \gamma L$ [15]. This is the origin of the notation 'angular momentum compensation point'.

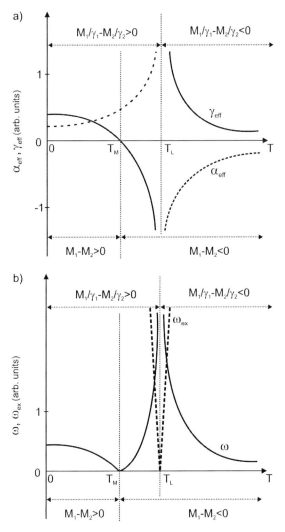

Figure 2.10: a) Possible scheme of the dependence of α_{eff} (dashed lines) and γ_{eff} (solid lines) on the temperature T. α_{eff} and γ_{eff} diverge at T_{L}. γ_{eff} is zero at T_{M}. b) Corresponding scheme of the ferromagnetic resonance frequency ω (solid lines) and exchange resonance frequency ω_{ex} (dashed lines) as function of T. ω is zero at T_{M} and diverges at T_{L}, ω_{ex} is zero at T_{L}. Note the steep slope of ω_{ex} in the vicinity of T_{L}.

H_c should diverge. This can be understood as follows: as the magnetization gets smaller towards the magnetization compensation point, it is getting more difficult to exert a torque $T = M \times H$ on the magnetization. At T_M the magnetization is $M = M_1 - M_2 = 0$ and an infinitely strong field would be required to move/switch the magnetization. Hence, it is not possible to alter the magnetic energy of the system. Note that T_M and T_L are expected to be very close together if the g-factors of both sublattices are nearly equal.

2.5 All-Optical Pump-Probe Schemes

Ultrafast all-optical pump-probe experiments are often used for the investigation of the electronic relaxation processes in metallic materials [59]. To describe the dynamical behavior of the systems, different models were developed. In the next subsections two phenomenological models will be introduced that are commonly used to describe electron, lattice and spin dynamics. Both models are based on heat baths for the different systems and the coupling between them. A short laser pulse is used to trigger the dynamics based on an excitation of electrons to a non Fermi-Dirac distribution. Note that 'temperature' is an equilibrium concept. No temperature can be defined until the corresponding systems are internally thermalized, i. e. until the distribution of the electron system/ phonon system can be fitted using a Fermi-Dirac / Bose- Einstein distribution [60].

2.5.1 Two Temperature Model

In the phenomenological two-temperature model (2-T model), originally proposed by Anisimov [61], the electron and phonon system are modelled as two heat baths with temperatures T_e and T_p, respectively. Here T_e is defined as the temperature of the electron system and T_p is the 'real' lattice temperature. Both systems are connected via electron-phonon interactions that are described by a parameter g_{ep}. The term phenomenological implies that the underlying physics is not directly visible in the model. At the beginning the baths are in equilibrium, i. e. $T_{e,i} = T_{p,i}$. In Fig. 2.11 a) the electron distribution for an initial temperature $T_{e,i} > 0$ K, which is based on a Fermi-Dirac distribution, is shown. Upon absorption of a short laser pulse electrons are excited from the occupied states below the Fermi level E_F to empty states above the Fermi level. This process can be considered to be instantaneous within the laser pulse duration, which

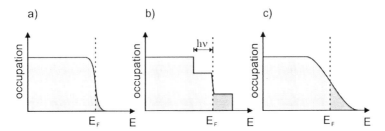

Figure 2.11: Schematic illustration of the electron distribution around the Fermi energy E_F. a) depicts the situation for an initial temperature $T_{e,i} > 0$ K before the pulse. b) After the perturbation by the laser pulse a strong non-equilibrium exists that does not allow to define a temperature for the electron system. c) After thermalization of the hot electrons a new equilibrium distribution with $T_{e,h} > T_{e,i}$ is established.

in our case is about 200 fs. A non-equilibrium distribution of highly energetic electrons, that are also referred to as hot electrons, is created (cp. Fig. 2.11 b)). For this situation no temperature can be defined. After a rapid thermalization of the hot electrons within some hundred femtoseconds a new equilibrium is reached and a description of the electronic system with an electronic tempera- ture is valid again (cp. Fig. 2.11 c)) [62]. The phonon system, i. e. the lattice, that can be characterized by a lattice temperature T_p, is initially unaffected by the optical excitation. However, through the coupling of the electron system to the phonon system by electron-phonon interactions both baths start to equili- brate. This leads to an increasing lattice temperature and a decreasing electron temperature until the thermal equilibrium is reached and the whole system can be described by a single temperature T.

The knowledge of the timescales on which the thermalization processes take place is desirable. A first approximation of the electron-electron (e-e) scattering time τ_{e-e} for a hot electron can be given by

$$\tau_{e-e}(E) = \tau_0 \left(\frac{1}{E - E_F} \right), \tag{2.32}$$

where τ_0 is a material specific constant and E_F is the Fermi energy [63]. Typical e-e-scattering rates are of the order of 10 fs for $E - E_F = 1$ eV, whereas the thermalization of the electron system takes few hundreds of femtoseconds. The times to equilibrate the electron and phonon system are ranging from < 0.5 ps for transition metal compounds [64] to several picoseconds for noble metals [65–68]. The different mechanisms that contribute to e-e-scattering and electron-phonon (e-p) scattering are well studied, both experimentally and theoretically [60, 65, 67–71].

A possible representation of the 2-T model is [65, 72]

$$C_e(T_e)\frac{\partial T_e}{\partial t} = \nabla \kappa_e \nabla T_e - g_{ep}(T_e - T_p) + P(\boldsymbol{r}, t), \qquad (2.33)$$

$$C_p(T_p)\frac{\partial T_p}{\partial t} = g_{ep}(T_e - T_p), \qquad (2.34)$$

with the electronic and phonon heat capacity C_e and C_p, respectively. κ_e is the electron thermal conductivity and g_{ep} is the coupling constant between the electron and phonon bath. The first term on the right hand side of equation (2.33) describes the diffusive electron heat transport out of the excited region, the second term the coupling of energy between the electron and phonon system. P describes the heating of the electrons due to the laser pulse. As the laser pulse only affects the electron system and the heat diffusion of the phonons is on larger time scales equation (2.34) only consists of the coupling term. The model is oversimplified in many aspects but nevertheless is often applied successfully [73].

2.5.2 Three Temperature Model

In the 2-T model the spin system is neglected. In order to model spin dynamics due to the perturbation by a laser pulse a new heat bath for the spin system is introduced. This model, that includes the 2-T model, is commonly named 3-T model. The spin temperature T_s is defined via the conventional M-T-diagram (cp. Fig. 2.12), i. e. the absolute value of $\boldsymbol{M}$ defines the spin temperature.

In Fig. 2.12 the influence of a short laser pulse on a sample with initial temperature T_1 is depicted. Due to the absorption of the laser beam the sample is heated and after thermalization the material is at a new equilibrium at the elevated temperature T_2. According to the classical $M(T)$-dependence [21] the final state exhibits a lower magnetization, therefore this process is referred to as demagnetization. The processes that are involved in the demagnetization process, and the time scales on which these processes are operative are a field of active research [74, 75].

The first attempt to include the spin system in a 3T-model was made by Vaterlaus et al. [76]. The approach of the 3-T model is to describe a magnetic state

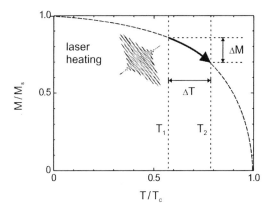

Figure 2.12: Schematic illustration of a typical M-T-diagram. Laser heating leads to a temperature increase of ΔT and a magnetization decrease of ΔM. The curve is used to define the spin temperature T_s.

by a single spin temperature. This is only a crude approximation, but often works well [73]. A common form of the 3-T model is

$$C_e(T_e)\frac{\partial T_e}{\partial t} = \nabla\kappa_e\nabla T_e - g_{ep}(T_e - T_p) - g_{es}(T_e - T_s) + P(\boldsymbol{r}, t), \quad (2.35)$$

$$C_p(T_p)\frac{\partial T_p}{\partial t} = g_{ep}(T_e - T_p) - g_{ps}(T_p - T_s), \quad (2.36)$$

$$C_s(T_s)\frac{\partial T_s}{\partial t} = g_{es}(T_e - T_s) + g_{ps}(T_p - T_s) \quad (2.37)$$

with the parameters analogously defined to the ones in the 2T-model. Hence, g_{es}, g_{ps} is the coupling constant between the electron, phonon and spin system, respectively. C_s is heat capacity of the spin system. A result of this model in the work of Beaurepaire [59] on polycrystalline Ni is shown in Fig. 2.13. The equations can be modified to achieve a more realistic description for a certain problem. For example, κ_e can often be neglected for the description of thin films [59, 72]. A schematic diagram which illustrates the dependencies between the baths is shown in Fig. 2.14. Remember that only the electron system is directly affected by the laser. Subsequent heating and equilibration of the other

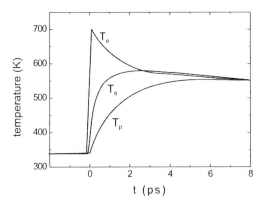

Figure 2.13: An example of fitted temperatures according to the 3-T-model after [59].

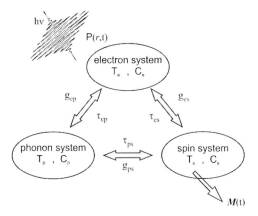

Figure 2.14: Interaction scheme of the 3-T model. The laser pulse only affects the electron system and leads to a nearly instantaneous increase of T_e. The subsequent energy transfer between the different baths is governed by the coupling constants g_{xy} and happens on timescales τ_{xy}.

systems is caused by interactions between the three heat baths.

At the beginning of the 1990s Vaterlaus et al. were able to determine a spin-lattice relaxation time $\tau_{ps} = 100 \pm 80$ ps for Gd by spin-polarized photoemission experiments [76, 77] in agreement with theoretical estimates by Hübner et al. [78]. In 1996 surprising results by Beaurepaire [59] using an all-optical pump-probe setup revealed a demagnetization time and an electron-spin relaxation time $\tau_{es} < 1$ ps, respectively. This time scale for the electron-spin relaxation has been confirmed for various materials, various measurement techniques and theoretical models [79–87]. The fact that the demagnetizing time is much shorter than the spin-lattice relaxation time was very surprising. The Einstein-de Haas effect [88] describes that for a change of the magnetization a transfer of angular momentum from the spin system to the lattice is required to fulfill the angular momentum conservation. Hence τ_{ep} was expected to be similar to the demagnetizing time τ_{es}. The underlying mechanism and interactions are still discussed controversially [73,74]. For further details the reader is referred to [73].

In this thesis we focus on magnetic processes that occur more than 10 ps after the pulse. We are not interested in the details of the excitation process itself. Optical artifacts for the detection of magnetization dynamics by time resolved magneto optic Kerr effect (TR-MOKE, cp. subsection 4.3.2) like bleaching effects [64,73] are not relevant on that time scale.

2.5.3 Excitation in All-Optical Pump-Probe Experiments

In all-optical pump-probe experiments the magnetization dynamics is directly triggered by the pump beam. Different mechanisms can cause the magnetization to be displaced out of its equilibrium. The processes, which are important for this thesis, will be presented here.

The excitation has to create a non-parallel alignment of the effective field $\boldsymbol{H}_{eff}$ and the magnetization $\boldsymbol{M}$ as according to the LLG-equation (2.14) a torque $\boldsymbol{M} \times \boldsymbol{H}_{eff} \neq 0$ is needed to trigger magnetization dynamics. As shown in subsection 2.5.1 the sample is heated locally by an intense laser pulse. This is the origin for all processes that can lead to the sufficient displacement of $\boldsymbol{M}$ and $\boldsymbol{H}_{eff}$. In most cases a canted magnetic state is required as it has a non-vanishing stray field. Thus the following discussions all are based on a canted state. For in-plane magnetized samples a canted state can be prepared by applying an adequate magnetic field perpendicular to the film surface.

In canted magnetic configurations the magnetization has an in-plane and an out-of-plane component, and thus a non-vanishing stray field. If M is shortened due to the heating of the excitation pulse, the magnitude of the stray field is also drastically altered (reduced). Hence H_{eff} moves and is not collinear to M anymore which triggers the magnetic precession. The change of magnetic anisotropies, which are temperature dependent as well, can have the same effect like the shortening of M on the stray field. A third possible contribution is a non homogenous heating throughout the sample (laterally as well as across the film thickness). This can additionally shift H_{eff} as the stray field is also modified. Note that various excitation mechanisms can be operative in a sample at the same time. A scheme of the excitation of a canted magnetic state and the subsequent magnetization dynamics is depicted in Fig. 2.15.

For all presented excitation processes the temperature change, and thus the change of M in the pumped area plays an important role. The temperature increase is closely related to the amount of energy that is deposited in the sample by the pump pulse. A measure for this deposited energy is the fluence of the laser beam ρ. All fluences in this thesis were calculated using the relation

$$\rho = \frac{2E_{\text{p}}}{\pi d_{\text{f}}^2} \tag{2.38}$$

with the energy per pulse E_{p} and the focus diameter d_{f}. The focus diameters are determined using a scanning technique on a structured sample [89]. According to equation (2.38) it is possible to tune the the energy deposition and therefore the temperature change due to the pump pulse, by tuning the focus diameter and the pulse energy of the pump beam.

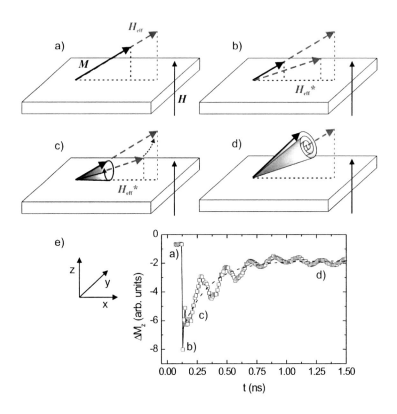

Figure 2.15: Illustration of a possible excitation scheme of magnetization dynamics in an all-optical pump-probe experiment. In a) M is aligned parallel to H_{eff} which corresponds to the equilibrium position without excitation. H depicts an applied out-of-plane field that prepares the canted magnetic state. b) shows the situation right after heating by the pump pulse. M is shortened due to the heating by the laser pulse and $H_{eff}*$ is the transient equilibrium position. Therefore M starts to precess around $H_{eff}*$ as sketched in c). As the system cools down $H_{eff}*$ approaches H_{eff} and M increases. After a certain time the the initial length of M and the initial H_{eff} is restored. M performs a damped oscillation according to the LLG-equation back to the initial equilibrium position as sketched in d). In e) a typical measurement on a 30 nm thick NiFe sample is plotted. The letters indicate the different states.

3 Samples

In this chapter the sample preparation and the properties of the samples will be presented. All samples that are discussed in the present thesis were sputter deposited using standard sputtering techniques except the CoGd gradient samples in chapter 7.

Rare-earth doped Permalloy and homogenous CoGd samples

The NiFe sample presented in chapter 5, the series of rare-earth doped NiFe samples presented in chapter 6 and the homogenous CoGd samples presented in chapter 7 were grown at the San Jose Research Center of Hitachi Global Storage Systems in California. These samples were sputter deposited by DC magnetron co-sputtering from single element targets. Glass or SiN-coated Si were used as substrates. For the CoGd samples a 200 nm thick Au or a 100 nm thick Al heatsink was deposited prior to the deposition of the CoGd films. Cap layers of Ta/Pt or Al_2O_3 in a thickness range from 3-5 nm were used to prevent oxidation of the films. An Ar-gas pressure of 3 mbar was used for sputtering. Film thicknesses for the CoGd were 30 nm or 50 nm, and the composition of the $Co_{1-x}Gd_x$ films was adjusted near $x = 0.22$ such that the magnetic compensation temperature T_M of the ferrimagnet was not too far away from room temperature (cp. Fig. 2.8 and the results by Hansen et al. [90]). The film thicknesses for the doped NiFe sample series was 30 nm. Sample thickness and composition were verified by X-ray reflectivity measurements and Rutherford backscattering (RBS) [91], respectively. The samples are all amorphous or polycrystalline with negligible anisotropies (≤ 1.7 kA/m (25 Oe)) (cp. Appendix A). Detailed information about the sample properties (compositions, thicknesses, M_s etc.) will be given in chapters 5,6.

CoGd gradient samples

The CoGd gradient samples presented in chapter 7 were prepared at the Dalhousie University in Halifax, Canada, using a special sputtering technique.

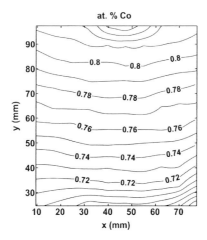

Figure 3.1: The concentration profile of the CoGd gradient sample on GaAs in atomic percent of Co. The linear gradient is oriented along the y axis.

One set was grown on Si (111) with native oxide using DC-sputtering and provided a continuous concentration gradient across the wafer (1.43 percent/mm). The composition gradient was confirmed using X-ray analysis. A second gradient sample was grown on GaAs (100) with a concentration gradient reduced by roughly a factor of ten (0.16 percent/mm). The samples were grown by magnetron sputtering at a base pressure of $7 \cdot 10^{-8}$ mbar. Single element targets were used. The sputtering chamber is equipped with a water-cooled rotating substrate table. The chamber pressure was maintained at 3 mbar of argon during deposition. The sputtering targets were covered with masks designed to control the composition as a function of position on the wafer. Masks to give linearly varying deposition and constant deposition across the substrate were used.

In another run under the same conditions continuous gradient films were also deposited on Si (111) wafers. These films were used for X-ray diffraction and electron microprobe analysis [91] to check the sample thicknesses and to determine the structure and composition as a function of position across the film. The concentration map of the CoGd gradient sample on GaAs is shown in Fig. 3.1. Details of the complex DC-sputtering procedure and the analysis methods have

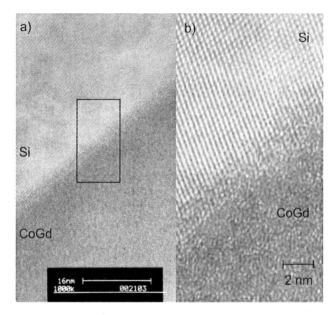

Figure 3.2: In a) a high resolution cross-sectional transmission electron microscope image of a CoGd gradient sample on Si is shown. In b) the area within the black frame is enlarged. The structured order of the Si substrate and the amorphous structure of the CoGd layer is evident. The overall thickness of the CoGd layer is 250 nm.

been previously described [92–94]. The film thickness was about 300 nm but varied around this value with the composition of the film. All films were capped with transparent 4 nm thick Al_2O_3 layers to prevent oxidation. For the experiments the wafers were cleaved into small pieces. Further detailed information about the sample properties will be given in chapter 7.

Based on the X-ray diffraction measurements all CoGd samples are amorphous. This is also confirmed using high resolution cross-sectional transmission electron microscopy (TEM) analysis. In Fig. 3.2 a cross-sectional TEM image is shown. The ordered structure of the Si substrate and the amorphous structure of the CoGd layer can be clearly distinguished.

The samples exhibit only very small magnetic anisotropies that can be neglected.

4 Experimental Techniques

In this chapter the experimental techniques that are used for the measurements of the magnetization dynamics will be introduced briefly. The basic equations that were used to extract the data in the experimental chapters will be discussed.

4.1 Ferromagnetic Resonance

Two different Ferromagnetic Resonance (FMR) techniques were used for the measurements of this thesis. The basic principles of FMR will be introduced in the following subsections.

4.1.1 Conventional Ferromagnetic Resonance

Conventional FMR is an established technique used to measure magnetization dynamics [95]. FMR occurs in the microwave range of frequencies. A sketch of our FMR setup is shown in Fig. 4.1.

A small sinusoidal microwave field with fixed frequency f is generated by a gun diode or klystron that is connected to a shorted waveguide. The magnetic sample is mounted at the shorted end of the waveguide which is placed between the pole pieces of a large electromagnet. The rf-field excites the magnetization of the sample and a variable, simultaneously applied static magnetic field H_{dc} allows one to sweep the sample across the resonance condition. At the resonance field H_{FMR} the microwave absorption of the magnetic sample reaches its maximum. The FMR signal is obtained by monitoring the microwave losses of the reflected microwaves as a function of the external field via a diode microwave detector. In order to enhance the signal to noise ratio (S/N) the external field is modulated using small additional coils and lock-in detection is used.

For the in-plane configuration the external dc-field and the magnetization are oriented in the magnetic film plane, the resonance condition is [40,96]

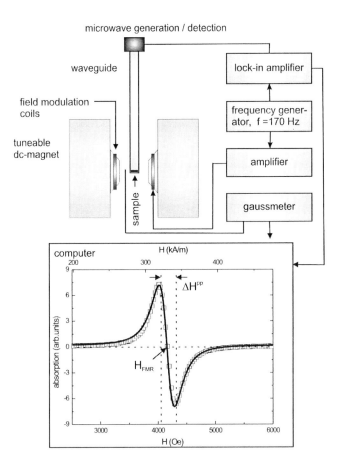

Figure 4.1: Scheme of the FMR setup. The modulation of the dc-field by the modulation coils allows for lock-in detection of the microwave absorption. A typical FMR measurement on a 30 nm thick NiFeHo2 sample (open circles) at 22.1 GHz and the corresponding fit (straight line) according to equation (4.6) are shown. The peak-to-peak linewidth H^{pp} and the position of the resonance field H_{FMR} are depicted.

$$\left(\frac{\omega}{\gamma}\right)^2 = \mu_0\,\mathcal{B}_{\text{eff}}\,\mathcal{H}_{\text{eff}}\Big|_{H_{\text{FMR}}}, \tag{4.1}$$

with $\omega = 2\pi f$, the gyromagnetic ratio γ, the effective magnetic induction $\mathcal{B}_{\text{eff}}$ and the effective magnetic field $\mathcal{H}_{\text{eff}}$. For the samples used within this thesis magnetic anisotropies can be ignored which leads to $\mathcal{B}_{\text{eff}} = \mu_0(H_{\text{dc}} + M)$ and $\mathcal{H}_{\text{eff}} = H_{\text{dc}}$. For the saturated case the following relation for the imaginary part of the rf-susceptibility $\chi = m^{\text{rf}}/h^{\text{rf}}$ with the rf-driving magnetic field $\boldsymbol{h}^{\text{rf}}$ and the rf-magnetization contribution $\boldsymbol{m}^{\text{rf}}$ is valid [40]:

$$\text{Im}[\chi] = M_{\text{s}}\frac{\mathcal{B}_{\text{eff}}}{\mathcal{B}_{\text{eff}} + \mu_0\mathcal{H}_{\text{eff}}}\Bigg|_{H_{\text{FMR}}}\frac{\Delta H}{\Delta H^2 + (H_{\text{dc}} - H_{\text{FMR}})^2}, \tag{4.2}$$

$$\Delta H = \alpha\frac{\omega}{\gamma\mu_0}, \tag{4.3}$$

where ΔH is the half width at half maximum (HWHM) linewidth. Thus the imaginary part of the rf-susceptibility is given by an almost perfect Lorentzian function. Note that these expressions are only valid for a fully saturated magnetic sample and therefore an uniform precession of the magnetization. As the external field is modulated for the lock-in detection, the FMR spectrum is proportional to the field derivative of the imaginary part of the microwave susceptibility (4.2).

In the perpendicular configuration the magnetization and the applied static magnetic field are both oriented perpendicular to the sample plane. Neglecting anisotropies the resonance condition is given by the following relation [40, 97]

$$\frac{\omega}{\gamma} = \mu_0(H_{\text{FMR}} - M_{\text{s}}) \tag{4.4}$$

A typical FMR spectrum contains two important pieces of information, the line position and the linewidth (cp. Fig. 4.1). The line position is determined by the internal fields, and its angular and frequency dependence provides information e. g. about anisotropies and the g-factor. The linewidth is related to the magnetic damping as can be seen in equation (4.3). Its angular and frequency dependence provides insight into the magnetic damping mechanisms. The linewidth and line position are extracted by fitting the FMR data to the derivative of a symmetric Lorentzian. The absorption function A is given by [40]

$$A \sim \frac{\Delta H}{\Delta H^2 + (H - H_{\mathrm{FMR}})^2},\tag{4.5}$$

and the function to fit the FMR data is its derivative with respect to H

$$\frac{dA}{dH} \sim -\frac{2(H - H_{\mathrm{FMR}})\Delta H}{[\Delta H^2 + (H - H_{\mathrm{FMR}})^2]^2},\tag{4.6}$$

A sketch of our FMR setup and measured absorption line with the corresponding fit according to equation (4.6) are shown in Fig. 4.1. The resonance field H_{FMR} is given by the zero crossing and the peak-to-peak linewidth H^{PP} is equal to the distance between the inflection points. The peak-to-peak linewidth H^{PP} is related to the HWHM linewidth ΔH by

$$\Delta H = \frac{\sqrt{3}}{2}\Delta H^{\mathrm{PP}}\tag{4.7}$$

In our setup it is possible to tune the sample temperature in a range from 350 K to 400 K with an accuracy of 3 K in a dc-field up to $1.3 \cdot 10^6$ A/m (16 kOe). Different microwave sources allow measurements of microwave frequencies of up to 36 GHz.

Beside the uniform mode also PSSW modes can be detected in FMR experiments. It is only possible to excite odd PSSW (p=1, 3, 5,...) by a homogenous rf-field as only those waves have a non-vanishing interaction with the rf-field [30,98]. Therefore only odd PSSW can be detected in our ferromagnetic resonance experiments.

Separation of intrinsic and extrinsic damping by FMR
Intrinsic damping causes a resonance linewidth that is linearly proportional to the microwave frequency (cp. eqn. (4.3)) [40]. However, in experiments often a linear frequency dependence with an extrapolated non-zero linewidth for zero frequency, i. e. $\Delta H(0) \neq 0$ is found [99]. Hence, the linewidth as a function of frequency is often interpreted using the simple relation

$$\Delta H(\omega) = \Delta H(0) + \alpha \frac{\omega}{\mu_0 \gamma},\tag{4.8}$$

where the linear term is assumed to be a measure of the intrinsic damping and
the magnitude of $\Delta H(0)$ depends on the sample quality. $\Delta H(0)$ is expected to
tend to zero for high quality samples and should therefore be a measure for the
extrinsic damping as this is caused by sample defects and inhomogeneities. In the
perpendicular configuration two-magnon scattering is switched off (cp. subsections 2.3.1 and 2.3.2). This is due the fact that the spin wave dispersion depends
on the angle of the magnetization θ_M. More details can be found in [32, 40]. If
FMR measurements are carried out in the in-plane and in the perpendicular
geometry, the difference of the obtained damping parameters can be a measure
for the strength of two magnon scattering.

4.1.2 Vector Network Analyzer Ferromagnetic Resonance

In contrast to conventional FMR in the novel technique of Vector Network Analyzer Ferromagnetic Resonance (VNA-FMR) [100, 101] the external static magnetic field is kept at a fixed value and the frequency of the microwave radiation is
tuned. The source and detection unit for the radiation is a commercial network
analyzer (Agilent PNA E8362A) which is connected to a high bandwidth waveguide. The VNA sweeps the frequency in a range from 45 MHz to 20 GHz. As in
FMR the magnetic sample is excited by the microwave field. At the resonance
frequency energy is transferred to the sample and a reduced transmitted signal
amplitude can be detected by the VNA. Note that this method does not affect
the domain structure of the magnetic sample in contrast to FMR if no external
field is used. In VNA-FMR no lock-in detection like in conventional FMR is
used. Therefore the resonance peak is described by a Lorentzian,

$$y = \frac{2A}{\pi} \frac{w}{(f - f_0)^2 + w^2},\tag{4.9}$$

where f_0 is the resonance frequency, w the HWHM linewidth and A the area
enclosed by the peak.

A sketch of the setup and a typical spectrum with a fit according to equation
(4.9) are shown in Fig. 4.2. The coplanar waveguide is connected to the VNA
using microwave coaxial cables and hf-probes. An external field of up to 96 kA/m
(1.2 kOe) can be applied in the sample plane. For more detailed information
about the setup the reader is referred to [102, 103]. For samples with in-plane
magnetization the damping parameter α can be calculated using

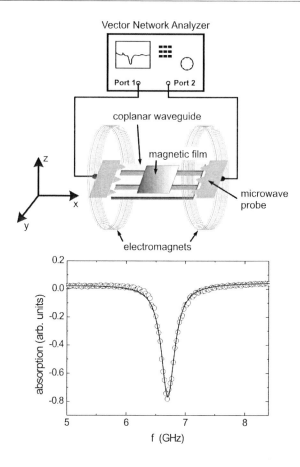

Figure 4.2: VNA-setup: The VNA acts as source and detector of the microwave radiation. It is connected to the waveguide by microwave compatible coaxial cables and probes. The sample is placed on top of the waveguide. In-plane fields can be applied using an electromagnet [102]. A typical signal of a 30 nm thick NiFe sample (open circles) in an in-plane magnetic field of 37.4 kA/m (470 Oe) and the corresponding fit according to equation (4.9) are shown in the lower panel.

$$\alpha = \frac{2\Delta\omega}{\gamma(\mathcal{B}_{\text{eff}} + \mu_0\mathcal{H}_{\text{eff}})}, \tag{4.10}$$

with the HWHM of the frequency linewidth $\Delta\omega$ [102,103]. $\mathcal{B}_{\text{eff}}$ and $\mathcal{H}_{\text{eff}}$ contain possible contributions of anisotropies and the applied magnetic field. They equal the relations of the conventional FMR (cp. subsection 4.1.1). Hence, in our case equation (4.10) is

$$\alpha = \frac{2\Delta\omega}{\mu_0\gamma(M + 2H_{\text{dc}})}. \tag{4.11}$$

4.2 Brillouin Light Scattering

Brillouin Light Scattering (BLS) is a spectroscopic method for the investigation of magnetic excitations using inelastic scattering of light. It is based on interactions between single photons with energy $\hbar\omega_i$ and momentum $\hbar\boldsymbol{k}_i$ and magnons, the quanta of spin waves, with energy $\hbar\omega$ and momentum $\hbar\boldsymbol{k}$ [35]. The interactions satisfy the energy and momentum conservation which can be written as

$$\hbar\omega_s = \hbar(\omega_i \pm \omega), \tag{4.12}$$

$$\hbar\boldsymbol{k}_s = \hbar(\boldsymbol{k}_i \pm \boldsymbol{k}), \tag{4.13}$$

where $\hbar\omega_s$ and $\hbar\boldsymbol{k}_s$ are the energy and the momentum of the scattered photon, respectively. This is schematically illustrated in Fig. 4.3. The scattered photon gains (loses) energy and momentum if a magnon is annihilated (created) during the scattering process. Equation(4.13) reveals that the wave vector $\boldsymbol{k}$ of the spin wave equals the wave vector $\boldsymbol{k}_s \mp \boldsymbol{k}_i$. In the case of a frequency increase (reduction) of the scattered photon the process is called Anti-Stokes (Stokes) process. For finite temperatures ($T \gg \hbar\omega/k_{\text{B}} \approx 1$ K) both processes have about the same probability. In a classical treatment the scattering can be understood as follows: Due to magneto-optical effects a phase grating is created in the material by the spin wave. It propagates with the phase velocity of the spin wave. Light is Bragg-reflected from the phase grating with its frequency Doppler shifted by the spin wave frequency [35].

A typical BLS setup consists of a laser beam with fixed wave length. The wave vector $\boldsymbol{k}$ can be adjusted by the scattering geometry [35] allowing to determine

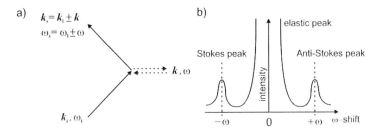

Figure 4.3: a) Scheme of the scattering process between a photon with energy $\hbar\omega_i$ and momentum $\hbar\,\boldsymbol{k}_i$, and a spin wave with energy $\hbar\omega$ and momentum $\hbar\,\boldsymbol{k}$. b) Corresponding schematic BLS spectrum.

the spin wave dispersion $\omega(\boldsymbol{k})$. In micro-focus BLS a spatial resolution of a few hundred nm can be achieved due to focussing of the beam. In modern BLS spectrometers the frequency shift of the light is analyzed using tandem Fabry-Perot spectrometers and the frequency resolution is about 0.2 GHz [104–107]. The merits of BLS lie in the large frequency range (up to 100 GHz and above) and the possibility to investigate thermally excited modes. In addition to thermally activated spin waves, it is also possible to measure spin waves, that are externally excited e. g. via a micro-coil [108]. BLS is reviewed in detail in [31,35,104,109]. The frequency of the spin wave can be directly extracted from the measured frequency spectrum. For samples with in-plane magnetization the damping parameter α can be calculated using equation (4.10) as BLS and (VNA-) FMR both are frequency domain techniques [102].

4.3 Time Resolved Magneto Optic Kerr Effect

4.3.1 The Magneto Optic Kerr Effect

The magneto optic Kerr effect (MOKE) [110, 111] has been an important and very useful tool for the investigation of magnetic samples for a long time [19]. The effect causes the plane of polarization of linearly polarized light to rotate slightly and become elliptically polarized when reflected by a magnetic sample. MOKE is a quantum mechanical phenomenon [112] but for the basic understanding a classical picture in terms of the Lorentz model is provided here [19].

This shall be explained for the so-called polar geometry which is displayed in Fig. 4.4 [1]. The electrical field E of the incoming light causes the electrons in the magnetic material to oscillate along the electrical field direction. Due to the magnetization M the electrons are additionally deflected by the Lorentz force which creates an additional oscillatory Kerr component $v_{lor} \propto E \times M$ perpendicular to the primary motion and the direction of the magnetization. This component leads to the emission of linear polarized light which has a different polarization plane with respect to the incoming light. This light is superimposed to the regularly reflected light that is polarized in the plane of the incident light. Therefore, the plane of polarization of the reflected light is slightly rotated and the beam becomes elliptically polarized.

There are three principle configurations for Kerr effect measurements (cp. Fig. 4.4) that differ in the relative orientation between the plane of incidence to the magnetization and the sample plane. In the following ψ denotes the angle of the incoming beam with respect to the sample normal (cp. Fig. 4.4):

- Polar MOKE: The magnetization M is oriented perpendicular to the sample surface. For $\psi = 0$ the Kerr rotation is maximal. The effect is independent of the incoming polarization direction.

- Longitudinal MOKE: The magnetization M lies along the plane of incidence and parallel to the surface. Only for $\psi \neq 0$ there is a resulting rotation of the reflected beam. A change of the polarization of the incoming light from parallel to perpendicular leads to an opposite sense of rotation.

- Transverse MOKE: The magnetization M lies perpendicular to the plane of incidence and parallel to the sample surface. For parallel polarized light a change in the intensity of the reflected light instead a rotation of the polarization occurs.

The polar MOKE is roughly ten times bigger than the longitudinal one [40]. The Kerr rotation is usually in the range of $10^{-3} - 10^{-5}$ rad [113]. Note that for arbitrary directions of M there may be a mixing of the different magneto optic effects.

The light reflected of the magnetic sample only interacts with a thin surface

---

[1]Similar arguments apply to the other Kerr configurations.

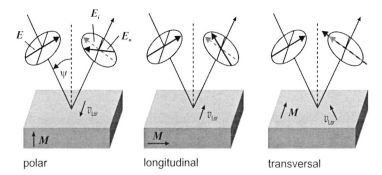

polar   longitudinal   transversal

Figure 4.4: Sketch of the three principle configurations for MOKE. M indicates the direction of the magnetization, E the polarization direction of the incoming and E_n the polarization direction of the reflected beam. E_i depicts the initial polarization direction. The rotation is caused by the additionally component $v_{lor} \propto E \times M$ due to the Lorentz force that is exerted on the moving electrons inside the magnetic medium.

layer that is determined by the skin depth. The skin depth for visible light is of the order of some 10 nm for metals [19,114]. Therefore MOKE can only provide information about the magnetization in this thin layer. For further details the reader is referred to [19].

4.3.2 Time Resolved MOKE

For the investigation of the magnetization dynamics using MOKE two important features compared to static MOKE setups are necessary. First, one has to use a pulsed laser source to achieve the temporal resolution and second, one has to perform a stroboscopic experiment in order to enhance the signal to noise ratio. An introduction and overview of time resolved MOKE experiments can be found in [115]. The spin dynamics can be triggered using various methods. The pump pulse can create a magnetic field pulse for example via a combination of a micro-coil and a photoswitch or a fast photodiode [116,117]. Another possibility is driving current directly through photo absorption in a semiconductor underlayer [118]. The magnetic field that is connected to the photo current excites the magnetization dynamics. For some materials it is possible to use the inverse Faraday effect to trigger the dynamics [119]. A further commonly used technique is to disturb the magnetization by sufficient heating of the spin system due to absorption of the pump pulse [59, 84, 120]. This technique is often referred to as all-optical. This approach will be followed in the presented experiments. The excitation scheme for our samples has been discussed in subsection 2.5.3.

To measure the temporal evolution of the magnetization after the excitation by a pump pulse, a probe pulse with variable time delay with respect to the pump pulse is used. Hence, one speaks of a pump-probe experiment. Through the evaluation of the Kerr signal of the probe pulse as function of the time delay between pump and probe pulse, information about the magnetization as function of time is obtained. The TR-MOKE setup is schematically depicted in Fig. 4.5.

A commercially available Ti:Sapphire Laser Mira 900, Coherent Inc., (repetition rate 80 MHz, $\sim$ 150 fs pulse length, central wave length of 840 nm, pulse energy $\sim$ 1 nJ) generates a pulse train that seeds the regenerative amplifier (RegA 9000, Coherent Inc., repetition rate 10-300 kHz, $\sim$ 150-250 fs pulse length, pulse energy $\sim$ 1-4 μJ). The 840 nm laser beam is frequency doubled using a barium beta borate (BBO) crystal due to second harmonic generation (SHG) [121]. After SHG the laser beam consists of two partial beams with wave lengths of 840 nm and 420 nm, respectively[2]. These two beams are separated using a dichroic beamsplitter. The red beam is used as pump, the blue beam as probe in the experiment. The length of the blue beam path is fixed whereas the red beam is guided via an optical delay stage which allows one to adjust the length of the red beam path. This adjustment is used to set the time delay between pump and probe pulses. The maximum delay time is 2.5 ns. The resolution of the delay stage provides sub-picosecond temporal resolution. The red beam is chopped at a frequency of approximately 1 kHz using a chopper wheel. This modulation of the excitation allows one to use lock-in detection. A scanning mirror unit is used to overlap the two beams on the sample. The red beam is focussed by an objective or a lens and strikes the sample perpendicularly. The blue beam passes a polarizer and an objective or a lens before it strikes the sample under an angle of approximately 45° with respect to the sample normal. The reflected blue beam passes an interference filter in order to suppress stray light. A Wollaston prism splits the light into two orthogonal polarization components. These components are measured via a balanced two diode detector which provides the difference

---

[2]in the following the 840 nm beam will be referred to as red beam, the 420 nm beam as blue beam.

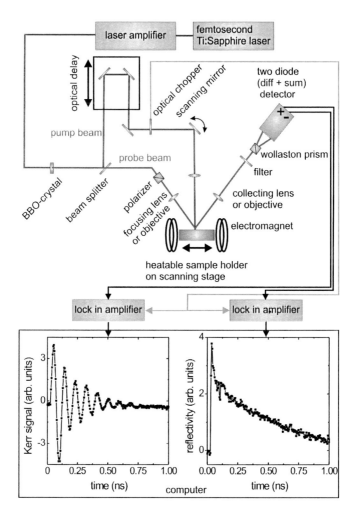

Figure 4.5: Scheme of the TR-MOKE setup: In the all-optical approach the 840 nm pump beam triggers the magnetization dynamics, the (frequency doubled) 420 nm probe beam detects the transient magnetization and reflectivity (an example of a CoGd sample is shown). The scanning mirror allows for exact overlapping of both beams. The time delay between the fs-pulses for the excitation and detection is set via a variable optical delay. An optical chopper modulates the pump beam to enable lock-in amplification.

and the sum of the two components. The signals are analyzed via two lock-in amplifiers that are locked on the reference frequency of the chopper. Under proper conditions [64,73] the difference signal is proportional to the Kerr signal and the sum signal is proportional to the reflectivity of the sample, which is a measure for the temperature of the probed area [122]. Note that the measured signals provide no absolute values. The signals are proportional to the change of the measured quantity i. e. in the case of the Kerr measurements the signal at a certain time t is proportional to $|M(t) - M(0)|$ with M(0) representing the magnetization in equilibrium. Hence, the signals can be interpreted as the difference between the signal with and without pump pulse.

There are mainly two reasons why two different wavelengths are used in the experiment. First the spatial resolution is higher for light with lower wave length as the minimal focus diameter for an objective is determined by

$$d_{\mathrm{f}} = 1.22 \frac{\lambda}{N} \tag{4.14}$$

with the wave length of the light λ and the numerical aperture of the objective N [114]. In our case N is 0.2 which allows minimal focus diameters of approximately 3 µm. The lenses used instead of the objectives have a focal length of 75 mm which allow minimal foci of approximately 15 µm [89].

The second reason is that one can use a filter that only transmits the blue beam in front of the detector. This excludes effects by the detection of stray light.

The time resolution in our experiment is limited by the length of the laser pulses to approximately 150-250 fs. The samples are mounted on a translation stage with spatial resolution of 1 µm. Furthermore, it is possible to heat the samples up to 400 K with a temperature stability better than 1 K. The electromagnet generates magnetic fields up to 143.2 kA/m (1.8 kOe). For further details on parts of the setup see [89].

4.3.3 Evaluation of the TR-MOKE Data

The goal of the data evaluation is to obtain information about the precessional frequency and the damping parameter α of a magnetic sample. The LLG-equation (2.14) can be solved analytically for small angle motions. We assume the magnetization and the applied magnetic field to lie in-plane along the x-axis. We further assume that the small angle approximation ($M_x \approx M_{\mathrm{s}} = const.$) is fulfilled. Then the LLG-equation is given by a set of two coupled differential equations in m_y and m_z, where m_y and m_z are the y- and z-component of the rf-magnetization m, and m_z is perpendicular to the sample plane [40]. The coupled equations can be transformed in a second order differential equation for a damped harmonic oscillator in m_z, namely

$$(1 + \alpha^2)\ddot{m}_z + \gamma\alpha(\mathcal{B}_{\mathrm{eff}} + \mu_0\mathcal{H}_{\mathrm{eff}})\dot{m}_z + \gamma^2\mu_0\mathcal{H}_{\mathrm{eff}}\mathcal{B}_{\mathrm{eff}}m_z = 0, \qquad (4.15)$$

where the effective magnetic induction $\mathcal{B}_{\mathrm{eff}}$ and the effective field $\mathcal{H}_{\mathrm{eff}}$ contain magnetic anisotropies and the applied magnetic field. For all samples in this thesis the anisotropies are weak and can be neglected. Therefore $\mathcal{B}_{\mathrm{eff}} = \mu_0(H_{\mathrm{dc}} + M_s)$ and $\mathcal{H}_{\mathrm{eff}} = H_{\mathrm{dc}}$ (cp. section 4.1.1).

Using the ansatz $m_z = ce^{iAt}$ and considering $\alpha \ll 1$, one obtains

$$A = \frac{i\gamma\alpha(\mathcal{B}_{\mathrm{eff}} + \mu_0\mathcal{H}_{\mathrm{eff}})}{2} + \gamma\sqrt{\mu_0\mathcal{B}_{\mathrm{eff}}\mathcal{H}_{\mathrm{eff}}}. \qquad (4.16)$$

In the small damping limit the solution for the FMR precession is given by

$$m_z(t) = m_z(0)e^{-t/\tau}\cos(\omega t + \phi), \qquad (4.17)$$

where $\tau = 2/\alpha\gamma(\mathcal{B}_{\mathrm{eff}} + \mu_0\mathcal{H}_{\mathrm{eff}})$ is the decay time, $\omega = \gamma\sqrt{\mu_0\mathcal{B}_{\mathrm{eff}}\mathcal{H}_{\mathrm{eff}}}$ is the angular frequency of the precession and ϕ is the phase. The damping parameter α is therefore given by the relation

$$\alpha = \frac{2}{\tau\gamma\mu_0(2H_{\mathrm{dc}} + M_s)}. \qquad (4.18)$$

An example of TR-MOKE data and the corresponding fit according to equation (4.18) are shown in Fig. 4.6. Note that for magnetic fields that are applied in arbitrary directions or larger angle motion of the magnetization an analytical solution of the LLG-equation is usually not possible. Only numerical solutions can be found in these cases.

Beside the FMR mode is also possible to excite and detect PSSW in all-optical pump-probe experiments [120, 122].

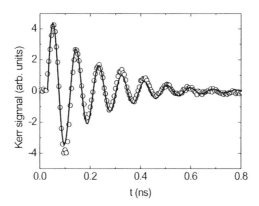

Figure 4.6: Example of a TR-MOKE measurement (open circles) on a piece of the CoGd gradient sample on Si in an applied magnetic in-plane field of 79.6 kA/m (1000 Oe). The solid line corresponds to a fit according to equation (4.17).

5 Experiments on NiFe

The goal of this chapter is to give an introduction into the magnetization dynamics experiments, with a focus on the TR-MOKE technique. A thin magnetic $Ni_{80}Fe_{20}$ film is used for the experiments. NiFe was chosen because of the simplicity of the system. It is a well studied ferromagnet and has a small intrinsic magnetic damping. The results of the various techniques will be discussed. The sample is a sputtered 30 nm thick NiFe film on a glass substrate. The film is capped with a 5 nm thick Ta layer in order to prevent it from oxidation. The experiments will be the starting point for investigations of NiFe samples with different rare-earth dopants in the next chapter.

5.1 Static Properties

First some basic magnetic properties of the used $Ni_{80}Fe_{20}$ will be presented. NiFe is a soft magnetic material, i. e. it has a small coercive field H_c. An easy axis hysteresis loop as measured by MOKE is shown in Fig. 5.1 a) with H_c=40 A/m (0.5 Oe). In Fig. 5.1 b) a hard axis loop is depicted from which the anisotropy field, i. e. the saturation field of the hard axis loop, is extracted. Here H_a=200 A/m (2.5 Oe). As H_a is very small, the anisotropy can be neglected for the NiFe film. The saturation magnetization M_s was determined by Vibrating Sample Magnetometery (VSM) to be $M_s = 807 \pm 30$ kA/m (807 ± 30 emu/cm^3). In the 30 nm thick NiFe film the magnetization is oriented in the sample plane.

5.2 TR-MOKE Measurements

Now we want to focus on the TR-MOKE experiment and on the magnetization dynamics. For the excitation of the magnetization in an all-optical pump-probe experiment a canted magnetic state is required (cp. subsection 2.5.3). As the magnetization in the NiFe film is in-plane, we use a sample holder with a built-in

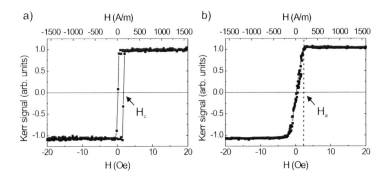

Figure 5.1: MOKE loops of $Ni_{80}Fe_{20}$. In a) an easy axis loop with $H_c=40$ A/m (0.5 Oe) and in b) a hard axis loop with $H_a=200$ A/m (2.5 Oe) is shown.

permanent magnet that produces a magnetic field perpendicular to the sample plane (cp. Fig. 5.2). The magnitude of the perpendicular magnetic field is about 159 kA/m (2 kOe). This holder will be referred to as the 'magnetic holder' in the thesis. Another holder that does not produce a magnetic field is used for some experiments. This holder will be referred to as 'non-magnetic holder'. Note that

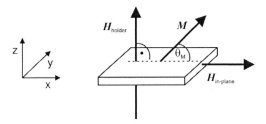

Figure 5.2: Canted magnetic state for the TR-MOKE measurements on NiFe. The perpendicular magnetic field of the magnetic sample holder H_{holder} tends to tilt the magnetization out-of-plane, an in-plane applied magnetic field $H_{in-plane}$ tends to align the magnetization in the sample plane. For $H_{in-plane} = 0$ the angle θ_M is about 11°. For 0 kA/m (0 Oe) $\leq H_{in-plane} \leq$ 103.5 kA/m (1300 Oe) the angle θ_M is approximately $10° \leq \theta_M \leq 11°$.

all results presented in this chapter were obtained using the magnetic holder. According to equation (2.22) the magnetization is tilted by 10°-11° with respect to the sample plane due to the perpendicular magnetic field of the sample holder and the accessible in-plane magnetic fields in our setup (cp. Fig. 5.2). For the maximum applied in-plane field of 103.5 kA/m (1300 Oe) the angle of the magnetization θ_M is 10°, for zero in-plane field 11°. Thus the change of the canting angle of the magnetization θ_M as a function of the applied in-plane magnetic field $\boldsymbol{H}_{in-plane}$ is small and will be neglected in the following.

The focus diameter of the blue probe beam and the red pump beam are 5 µm and 30 µm, respectively, in the TR-MOKE experiments. The pump fluence is 32 mJ/cm^2, if no different value is stated.

First the excitation of the magnetization via the pump pulse is discussed. In Fig. 5.3 a) the difference (diff) signal data of a TR-MOKE measurement with a time resolution of Δt=200 fs is plotted. The grey shaded area was also measured with a higher time resolution of Δt=50 fs (cp. Fig. 5.3 b)). In addition a laser pulse with 250 fs FWHM is depicted in Fig. 5.3 b). The pulse length in the experiment was determined using a commercial autocorrelator. The temporal pulse shape for our laser system can be calculated using

$$I(t) = \text{sech}^2 \left(\frac{1.76 \cdot t}{0.648 \cdot \Delta t_{\text{FWHM}}} \right), \qquad (5.1)$$

where $I(t)$ is the pulse intensity [123].

Prior to the excitation by the pump pulse the diff signal is constant. When the pump pulse arrives a sudden change in the signal occurs. The slope of the diff signal follows the intensity of the pulse. After about 500 fs the slope of the diff signal gets smaller, which seems to be correlated to the decaying pulse energy. It is well known that the diff signal includes magnetic as well as optical contributions such as the bleaching effect on ultra short time scales [64,73]. Hence an interpretation of the TR-MOKE data is difficult on that time scale. Our setup does not allow to obtain additional information about the magnetic system, e. g. the Kerr ellipticity. A changing ratio of the diff signal and the Kerr ellipticity can be an indicator that non-magnetic contributions are included in the signal. Therefore we did not perform more detailed investigations on ultra short time scales. Nevertheless, the overall shape of the diff signal trace can be understood in terms of the 3-T model (cp. subsection 2.5.2). After the

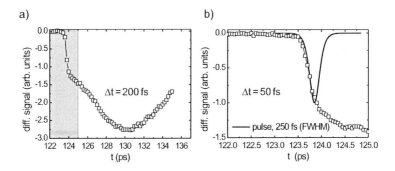

Figure 5.3: a) TR-MOKE difference signal of NiFe on short time scales with a time resolution $\Delta t = 200$ fs. The grey shaded area marks the time region for the data trace in b) which was obtained with a time resolution $\Delta t = 50$ fs. A laser pulse with 250 ps FWHM is also depicted in b). The diff signal follows the pulse shape. It takes about 6 ps for the difference signal to reach its minimum. The pump fluence is 32 mJ/cm^2.

instantaneous heating of the electron system by the laser pulse, the energy is subsequently transferred to the spin and lattice system, and the diff signal is reduced. The three heat baths equilibrate and approximately 6 ps after the arrival of the pulse the diff signal reaches its minimum. As the heat baths cool down the diff signal increases as the systems tends to its original equilibrium position. Note that we are interested in magnetization dynamics processes on the time scale about 10 ps after the pump pulse. We are not interested in the details of the the excitation process itself. On that time scale optical contributions to the diff signal can be neglected and the magnetic signal dominates [73]. Thus in the following parts of the thesis the diff signal will be referred to as Kerr signal.

In the following we will focus on the magnetization dynamics that occurs about 10 ps after the excitation.

A TR-MOKE measurement in an applied magnetic in-plane field of 23.9 kA/m (300 Oe) is shown in Fig. 5.4. The Kerr signal is plotted in Fig. 5.4 a) and the corresponding reflectivity signal is plotted in Fig. 5.4 b). In both signals a sharp jump is visible due to the pump pulse. As the time resolution is 10 ps,

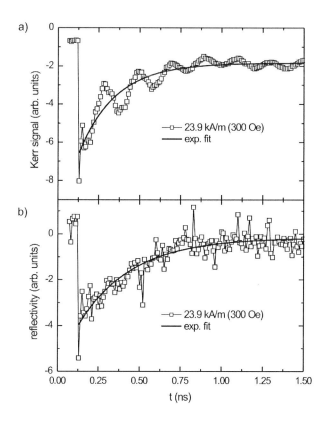

Figure 5.4: TR-MOKE measurements on NiFe. In a) the Kerr signal is plotted for an in-plane magnetic field of 23.9 kA/m (300 Oe). The precessional frequency is f=5.0 GHz In b) the corresponding reflectivity signal is shown. The solid lines correspond to fits according to equations (5.2) and (5.3). The decay times of the exponential background τ_b and of the reflectivity τ_r are $\tau_b = 0.24 \pm 0.05$ ns and $\tau_r = 0.34 \pm 0.05$ ns, respectively.

no details of the excitation are resolved. Note that in most of the presented measurements in this thesis time steps of 10 ps were used. An uniform precession of the magnetization on an exponential background can be seen in the Kerr signal. This is expected for our setup as we predominantly probe the out-of-plane component of the magnetization $\boldsymbol{M}$. The excitation and subsequent dynamics of the NiFe sample in the TR-MOKE measurement has been explained in detail in subsection 2.5.3 (cp. Fig. 2.15). The exponential background is due to the temperature dependence of $\boldsymbol{H}_{\text{eff}}*$ and the damped oscillations are triggered by the displacement of $\boldsymbol{M}$ and $\boldsymbol{H}_{\text{eff}}*$ and can be described by the LLG-equation. The decay time τ_{b} of the exponential background is determined using the relation

$$y(t) = A \cdot \exp(t/\tau_{\text{b}}) + y_0, \tag{5.2}$$

with the amplitude A, the decay constant τ_{b} and the offset y_0. The fit to the presented Kerr data is plotted in Fig. 5.4 a) and yields $\tau_{\text{b}} = 0.24 \pm 0.05$ ps .

The reflectivity is a measure for the temperature T of the probed area of the sample on the time scale ≥ 10 ps after the excitation [122]. This can be understood as follows: After the thermalization of the electron-, phonon- and spin-system to the same temperature, which takes about 10 ps, the temperature can be described by a single temperature T according to the 3-T model (cp. subsection 2.5.2). A change of the temperature T shifts the energy states and alters the occupation of the energy states. This leads to a change of the optical conductivity tensor and therefore of the reflectivity signal [122]. Thus the more the reflectivity signal $R(t)$ at a time t differs from the equilibrium reflectivity signal R(0) prior to the excitation, the higher is the temperature $T(t)$ compared to temperature $T(0)$. As can be seen in Fig. 5.4 b) the reflectivity, and consequently the temperature T of the investigated part of the sample, is also exponentially decaying after the excitation. The decay time τ_{r} is determined using

$$y(t) = A \cdot \exp(t/\tau_{\text{r}}) + y_0. \tag{5.3}$$

The fit to the presented reflectivity data is shown in Fig. 5.4 b). $\tau_{\text{r}} = 0.34 \pm 0.05$ ns is found. The temporal decay of the magnetic background τ_{b} is slightly faster than the decay of the reflectivity τ_{r}. This shows that at a slightly increased temperature the displacement of the transient position of the effective field $\boldsymbol{H}_{\text{eff}}*$ with respect to the equilibrium position of the effective field $\boldsymbol{H}_{\text{eff}}$ is very small.

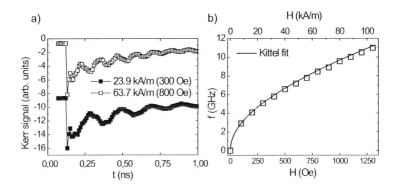

Figure 5.5: a) Kerr signal for an in-plane applied field of 23.9 kA/m (300 Oe) and 63.7 kA/m (800 Oe) with a precessional frequency of 5.0 GHz and 8.4 GHz, respectively. The traces are offset for clarity. b) Frequency (open squares) versus the applied magnetic in-plane field and corresponding Kittel fit (solid line) according to equation (2.17) with g=2.10, $M_x = 780$ emu/cm^3, $M_z = 160$ emu/cm^3, H_z =159.2 kA/m (2 kOe) and $N_z = 1$.

In the following we focus on the evaluation of the precessional frequency f and the damping parameter α of the uniform precession on the time scale of ≥ 10 ps. The precessional frequency f can be easily determined using a FFT. f depends on the applied magnetic field. In a larger applied magnetic field the precessional frequency f increases as expected for a magnetic signal (cp. Fig. 5.5 a)). In Fig. 5.5 b) the precessional frequency f is plotted as a function of the in-plane applied magnetic field. The solid line is a Kittel fit to the data according to equation (2.17).

The fit with the free parameters M_x, M_z, g and H_z yields the following parameters: $M_x = 780 \pm 30$ kA/m (780 ± 30 emu/cm^3), $M_z = 160 \pm 10$ kA/m (emu/cm^3) and H_z= 159 kA/m (2 kOe). From M_x and M_z the value of M_s is determined to be $M_s = 795 \pm 65$ kA/m (emu/cm^3) which in is agreement with $M_s = 807 \pm 30$ kA/m (783 emu/cm^3) as measured by VSM. Furthermore the canting angle θ_M can be calculated from the fit values of M_x and M_z to be 11°-12°, which is expected in our experiments (cp. Fig. 5.2). The extracted $g = 2.10 \pm 0.02$ agrees well with the g-factor of 2.120 ± 0.005 as determined by angle dependent

FMR (cp. Appendix A). Also the values for $H_{holder}=H_z=159.3$ kA/m (2 kOe) is reasonable.

Besides the precessional frequency f the damping parameter α can be extracted from the TR-MOKE measurements. To evaluate α from the data the magnetic background of the Kerr signal is removed by subtracting a running average value. A normalized gaussian window is used for the averaging. The width of the averaging window is chosen to be about one oscillation period of the magnetization to avoid oscillations in the running average. An example for the removal of the magnetic background is shown in Fig. 5.6 a). Note that it is sometimes also possible to remove the background by fitting the background according to equation (5.2). The data without background are fitted using an exponentially decaying cosine function as given in equation (4.17) (cp. Fig. 5.6 b)). Subsequently α is calculated using equation (4.18) since the canting angle θ_M is small. In Fig. 5.6 b) a fit with the decay time $\tau=0.488$ ns, which corresponds to a damping parameter $\alpha=0.0193$, is depicted. As equation (4.18) is only valid for in-plane magnetic fields and a pure in-plane magnetization the values for the damping parameter may have a systematic error. To verify the results for the damping parameter numerical solutions of the LLG-equation were performed. An example for $\alpha=0.019$ is shown in Fig. 5.6 b). A good agreement of the results for the damping parameter obtained by equation (4.18) and the numerical solution of the LLG-equation is found. Overall the numerical calculations yield $\alpha = 0.019 \pm 0.002$.

The decay time τ and the corresponding α according to equation (4.18) as a function of the in-plane applied magnetic field are summarized in Fig. 5.7. Both parameters are found to be not depending on the applied magnetic in-plane field. The decay time τ is evaluated to be 0.52 ± 0.08 ns and the damping parameter to be $\alpha = 0.018 \pm 0.003$. The damping parameter is about three times higher than published literature values of 0.007-0.008 [12, 124, 125] and values that are obtained using other techniques that will be presented later. The most likely reason for this discrepancy is an observed movement of the pump beam on the sample due to the mechanical time delay stage. This can significantly change the magnitude of the excitation and lead to a reduction of the decay time τ. Consequently a higher damping parameter is pretended in the TR-MOKE experiment. This is especially a problem for measurements on samples that have a low damping (small α) and therefore a large decay time τ, e. g. NiFe. For

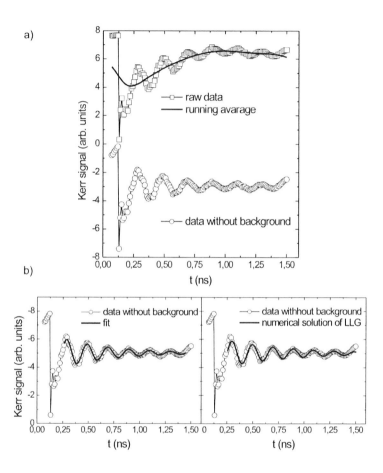

Figure 5.6: In a) the Kerr signal (open squares) of NiFe for an applied magnetic in-plane field of 23.9 kA/m (300 Oe) with a precessional frequency of 5.0 GHz is shown. The solid line corresponds to the running average of the data. In addition the data without background (open circles) are depicted. (The traces are offset for clarity). In b) a fit to the data without background according to equation (4.17) with the decay time $\tau = 0.488$ ns and the corresponding damping parameter $\alpha = 0.0194$ (left) and a numerical solution of the LLG-equation with $\alpha = 0.0190$ (right) are plotted.

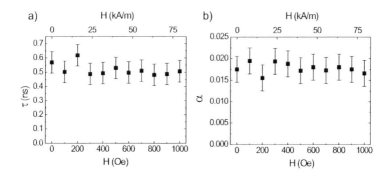

Figure 5.7: a) Decay time τ of the magnetic precession as a function of the applied magnetic in-plane field. The corresponding damping parameter α is shown in b). Both parameters are constant.

materials with larger magnetic damping the decay time will be smaller and the effect is less pronounced. This is the case for most of the RE-doped NiFe- and the CoGd-samples close to the compensation points that will be presented in the following chapters. Furthermore the error due to the stage movement is systematic and the same for all investigated samples. Hence relative statements about results as they will be given in chapters 6 and 7 are valid.

Another contribution to the increased damping in the TR-MOKE experiment can be due to radial damping via spin waves with wave vectors $k \neq 0$ that are excited through inhomogeneities during the optical excitation.

Now measurements with different pump fluences are presented in Fig. 5.8 a). A clear power dependence can be recognized in the Kerr data. With decreasing pump power the Kerr signal and the exponential magnetic background decreases. This is expected as the excitation of the magnetization and the temperature of the probed area scale with the pump power. A detailed analysis of the precessional frequency f by means of FFT and a fitting procedure as described above reveals no significant change of f as a function of the laser fluence. Hence M_{s} is nearly constant for the different pump fluences on the time scale on which the precessional motion takes place. Otherwise the frequency would change

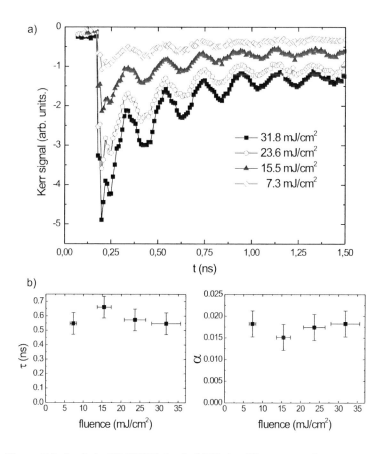

Figure 5.8: In a) the TR-MOKE signal of NiFe for different pump fluences in an applied magnetic in-plane field of 23.9 kA/m (300 Oe) is shown. The precessional frequency $f = 5.0 \pm 0.1$ GHz is found to be constant. In b) the decay time τ and the corresponding damping parameter α are plotted. Both parameters are also not dependent on the pump fluence.

(cp. eqn. (2.16)). The value of f is extracted to be $f = 5.0 \pm 0.1$ GHz. The decay time τ and the corresponding damping parameter α are plotted versus the pump fluence in Fig. 5.8 b). Both parameters are nearly constant within the examined pump power range. Therefore an arbitrary pump power could be selected for further measurements.

5.3 VNA-FMR and Conventional FMR Measurements

In order to verify the TR-MOKE results additional experiments using (VNA)-FMR were carried out on another piece of the same NiFe sample.

VNA-FMR measurements were performed as a function of the applied in-plane magnetic field. In Fig. 5.9 a typical VNA-FMR trace is shown. The frequency linewidth and the precessional frequency are extracted using a fit according to

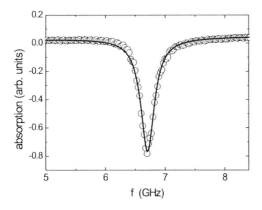

Figure 5.9: VNA-FMR data (open circles) for NiFe in a magnetic field of 37.4 kA/m (470 Oe). The fit (solid line) according to equation (4.9) yields a frequency linewidth of $\Delta f = 0.150 \pm 0.008$ GHz which corresponds to a damping parameter $\alpha = 0.0096 \pm 0.0005$.

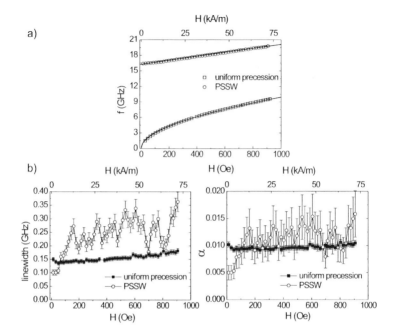

Figure 5.10: a) Measured frequencies of the uniform precession mode (open squares) and the first order PSSW (open circles). The solid lines correspond to fits according to equation (2.19) and (2.26, respectively, with $M_s = 810$ kA/m (emu/cm³), g= 2.12, $A = 8.8 \pm \cdot 10^{-12}$ J/m ($8.8 \cdot 10^{-7}$ erg/cm) and d=30 nm. In b) the frequency linewidth as function of the applied magnetic field is plotted for both modes. The corresponding damping parameters are shown in c).

equation (4.9). The damping parameter α is calculated using equation (4.11). Figure 5.10 summarizes the results obtained by VNA-FMR. Two modes could be identified in the data, the uniform precessional mode and the first order perpendicular standing spin wave (PSSW). Fig. 5.10 a) shows experimental data and the according fits using equation (2.19) for the uniform mode and equation (2.26) for the PSSW with p=1, respectively. The fits with the free parameters M_s, g, d and A and $k_\parallel = 0$ yield the following parameters: $M_s = 810 \pm 30$ kA/m

$(810 \pm 30$ emu/cm^3), g= 2.12 ± 0.02, $A = 8.8 \pm 0.5 \cdot 10^{-12}$ J/m $(8.8 \pm 0.5 \cdot 10^{-7}$ erg/cm) and d $= 30 \pm 0.3$ nm. The values of M_s and g are again in good agreement with the values previously found in the thesis, the thickness d of the sample is also well reproduced. The value for the exchange stiffness constant A is found to be in the range of published data of $1.3 \cdot 10^{-11}$ J/m $(1.3 \cdot 10^{-6}$ erg/cm) [126, 127]. In Fig. 5.10 b) the frequency linewidth and the corresponding damping parameter α are plotted as a function of the applied magnetic field. Note that the large error bars for the PSSW measurements are due to the very small signal amplitude compared to the signal of the uniform precession. For the evaluation of the damping parameter α of the PSSW the relation

$$\alpha = \frac{2\Delta\omega}{\gamma\mu_0(2(H_{dc} + 2k^2A/\mu_0 M_s) + M_s)} \tag{5.4}$$

is used. This formula can be derived from (4.10) if the exchange field contribution $k^2A/\mu_0 M_s$ [128] for the spin waves is considered in the effective magnetic induction $\mathcal{B}_{eff}$ and the effective magnetic field $\mathcal{H}_{eff}$. The mean value for the damping parameter α (frequency linewidth) is evaluated to be 0.010 ± 0.001 $(0.015 \pm 0.001$ GHz) for the uniform precession mode and 0.011 ± 0.003 $(0.25 \pm 0.06$ GHz) for the PSSW. The damping parameter for the PSSW is comparable to the one of the uniform precession. The slightly higher value is very likely due to small deviations of the sample thickness d. Even small variations of d can lead to a significant shift of the resonance frequency (cp. eqn. (2.26)). Therefore various resonance frequencies which are close together coexist in the sample. The corresponding resonance peaks add up and cause an artificial broadening of the measured frequency linewidth. Thus a higher damping parameter is extracted from the VNA-FMR experiment in the case of the PSSW. The damping parameters for both modes are again found to be slightly larger than published values of about 0.007-0.008. Possible contributions may be two magnon scattering processes, inhomogeneous excitation due to the wave guide and inhomogeneities in the sample [12, 124].

Besides the VNA-FMR experiments conventional in-plane FMR measurements at 21.9 GHz and 35.3 GHz were performed. In Fig. 5.11 a typical measurement at 21.9 GHz is shown. The field linewidth is determined using a fit according to equation (4.6), the corresponding damping parameter α is extracted using equation (4.3).

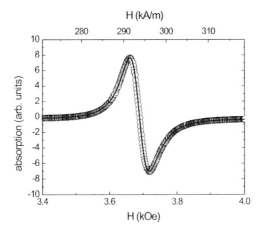

Figure 5.11: FMR data (open circles) for NiFe at a frequency of 21.9 GHz. The fit (solid line) according to equation (4.6) yields a field linewidth of $\Delta H = 4.46 \pm 400$ kA/m (56±5 Oe), which corresponds to a damping parameter $\alpha = 0.0076 \pm 0.0004$.

Fig. 5.12 a) depicts the field linewidth ΔH versus the frequency f. For frequencies below 10 GHz VNA-FMR measurements were performed. The frequency linewidth of the VNA-FMR was converted to the plotted field linewidth using $\Delta H_{\text{VNA}} = \alpha_{\text{VNA}} \omega / \gamma \mu_0$. The y-axis intercept of the straight line, which is a guide to the eye, lies at about 0.4 kA/m (5 Oe) in Fig. 5.12 a). According to equation (4.8) this implies that the extrinsic damping is very small in this case and that α is almost entirely intrinsic. Fig. 5.12 b) contains information about α as a function of the frequency. At low frequencies α increases drastically. The explanation for this behavior was given by Counil et al. [129]. Due to inhomogeneous excitation of the sample through the finite dimensions of the coplanar waveguide (CPW) spin waves are excited, which lead to a linewidth broadening of the uniform precession peak. In Fig. 5.12 c) a fit according to Counil's model is plotted. The used fit relation is [129]

$$\Delta\omega(k_{\max}) = \Delta\omega_{\text{int}}\sqrt{1 + \left(\frac{\omega_{\text{s}}(k_{\max}) - \omega_0}{\Delta\omega_{\text{int}}}\right)^2}, \qquad (5.5)$$

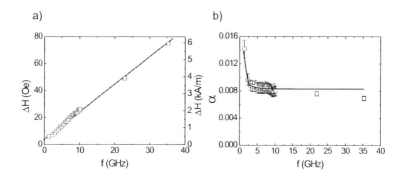

Figure 5.12: a) Field linewidth as a function of the FMR frequency. The straight line is a guide to the eye and reveals the expected linear dependence of ΔH vs. f according to equation (4.8). Results below 10 GHz are gained using VNA-FMR, above using FMR. b) depicts the evaluated damping parameter α vs. frequency f (open squares). α increases for small f. This can be explained by a model of Counil et al. [129]. The corresponding fit (solid line) according to equation (5.5) with $\alpha_0 = 0.0083$ and $k_{\mathrm{max}} = 2.5 \cdot 10^5 \mathrm{m}^{-1}$ is also plotted.

with the abbreviations

$$\omega_s^2(k_{\mathrm{max}}) = \omega_0^2 - \frac{1}{2}\gamma^2\mu_0^2 M_s(H_{\mathrm{dc}} - [H_{\mathrm{dc}} + M_s])k_{max}\,d, \tag{5.6}$$

$$\Delta\omega_{int} = \frac{1}{2}\alpha_0\gamma\mu_0 M_s\,, \tag{5.7}$$

$$\omega_0^2 = \gamma^2\mu_0^2 H_{\mathrm{dc}} M_s. \tag{5.8}$$

α_0 corresponds to the purely intrinsic part of the damping parameter, d is the sample thickness. k_{max} is the maximal wave vector for the excitations. For the calculation of k_{max} one has to account for non-uniform current densities in the CPW and the correlated inhomogeneities in the excitation fields [102,130]. The value of $k_{\mathrm{max}} \approx 2.7 \cdot 10^5 \mathrm{m}^{-1}$ is evaluated for the used 90 μm wide CPW [130]. The fit reproduces our data reasonably well. As usual the linewidth was transferred into the damping factor α by means of equation (4.11). Free parameters for the fit are α_0 and k_{max}. The fit which is shown in Fig. 5.12 yields $\alpha_0 = 0.0083$ and $k_{\mathrm{max}} = 2.5 \cdot 10^5 \mathrm{m}^{-1}$. The fit shows that α decreases with increasing frequency. This is also in agreement with the smaller values for α of the conventional FMR results at 21.9 GHz and 35.3 GHz, namely 0.0076 ± 0.0004.

This value agrees well with published data [12,124]. Furthermore the first order PSSW was found in the measurements. The damping parameter α is extracted to be 0.007 ± 0.001 according to equation (4.3). Thus the damping parameter of the uniform precession and the PSSW mode are the same.

Summarizing the measurements, it is reasonable to take $\alpha = 0.0076 \pm 0.0004$ as determined by FMR as the intrinsic damping parameter for NiFe. For quantitative measurements of the intrinsic damping parameter in VNA-FMR larger frequencies are required to exclude linewidth broadening effects due to spin wave excitations. Conventional FMR experiments provide reasonable high frequencies and are therefore better suited for the determination of the damping parameter. Note the lower values for the damping parameter α in Fig. 5.12 b) compared to the measurements presented in Fig. 5.10 b). Different parts of the pure NiFe sample are used for the measurements, therefore small deviations of the sample properties and of the excitation due to the CPW wave guide are very likely the reason for the smaller damping. A slightly altered (reduced) excitation can also be the reason that no signature of a PSSW was observed in the measurement presented in Fig. 5.12 as the excitation of PSSW modes is much weaker in VNA-FMR experiments than the excitation of the uniform precession.

5.4 Conclusion

It was demonstrated in this chapter that the use of pump-probe TR-MOKE and (VNA)-FMR techniques allow one the extensive characterization of the magnetization dynamics of NiFe.

The damping parameter in the TR-MOKE measurements was found to be $\alpha = 0.018 \pm 0.003$ which is in good agreement with numerical solutions of the LLG equation that yield 0.019 ± 0.002. In both cases the damping parameter is larger than the literature value of 0.007-0.008. This can be explained by a movement of the pump beam during the TR-MOKE experiment. The damping parameter α is constant as a function of the applied field and the laser fluence. In different VNA-FMR measurements the damping parameter was found to be in the range of 0.007-0.011. In the VNA-experiment the increased damping at low frequencies can be explained using the model by Counil et al. that considers inhomogeneous spin wave excitations due to the CPW. A small signature of a first order PSSW was found. The damping parameter for the first order

PSSW is $\alpha = 0.011 \pm 0.003$ and comparable to the one of the uniform precession 0.010 ± 0.001. Determinations of the linewidth as function of the frequency
show that the extrinsic damping contribution is small. The intrinsic damping
parameter α is evaluated by FMR to be $\alpha = 0.0076 \pm 0.0004$.

Due to measurements of the precessional frequency as a function of the applied
magnetic field the g-factor g= 2.10 ± 0.02 and the exchange stiffness constant
$A = 8.8 \cdot 10^{-12}$ J/m ($8.8 \cdot 10^{-7}$ erg/cm) were obtained and are in good agreement
to to published data and angle dependent FMR experiments.

6 Experiments on RE-doped NiFe

The previous chapter dealt with the magnetization dynamics of pure NiFe. Now the investigations of NiFe samples that are doped with various RE metals, namely Gd, Ho and Dy will be discussed. From the results by Reidy et al. [11], which were briefly summarized in the introduction, we expect the possibility to control the dynamic magnetic behavior by altering the RE dopant and RE content of the samples. All samples presented in this chapter are grown by co-sputtering on glass substrates. They are about 30 nm thick and capped with 5 nm Ta to prevent oxidation. The NiFeRE samples are ferrimagnets and hence the g-factor can be modified. A table that contains results for g and M_s obtained by angle dependent FMR and VSM, respectively, is given in Appendix A. All calculations in this chapter are carried out based on those data.

6.1 Magnetization Dynamics

Mainly VNA-FMR and FMR experiments were performed to get insight into the damping behavior of the different samples. The big advantage of these methods lies in the possibility to distinguish between intrinsic and extrinsic contributions to the damping parameter α. This is based on equation (4.8) in which the value of $\Delta H(0)$ mirrors the extrinsic contribution to the damping. The results are summarized in Fig. 6.1. The field linewidth is plotted as a function of the (VNA-) FMR frequency. Results below 10 GHz were obtained using VNA-FMR, results above 10 GHz using FMR. The conversion of the frequency linewidth from the VNA-experiments to the field linewidth is done analogously as presented in the previous chapter for Fig. 5.12. The solid lines are guides to the eye. In Fig. 6.1 a) NiFe, NiFeHo and NiFeDy samples are compared.

A clear increase of the magnetic damping for Ho is found with increasing Ho content as the slope of the lines is proportional to the damping parameter α by virtue of equation (4.8). Further the y-axis intercept is very small up to 4%

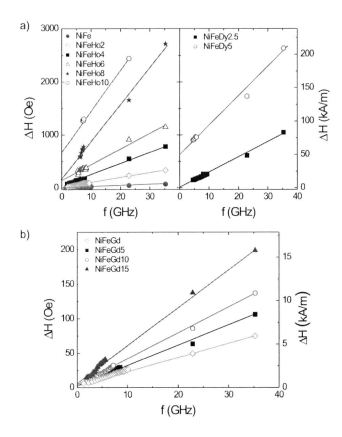

Figure 6.1: FMR results for different NiFe, NiFeHo, NiFeDy and NiFeGd samples. ΔH is plotted as a function of the FMR frequency f and RE content (results below 10 GHz are gained using VNA-FMR). The solid lines are guides to the eye. a) Increasing Ho and Dy contents lead to increasing damping, which is proportional to the slope of the solid lines according to equation (4.8). b) Adding Gd also increases the damping but the effect is much smaller (note the different y-axis scales).

Ho which reveals that the damping is mainly intrinsic. At the same time it is an indicator that the sample quality for low Ho doping is quite good as a main contribution to extrinsic damping is the two magnon scattering which is caused by sample defects and magnetic inhomogeneities. For 6%-10% Ho the extrinsic contribution seems to increase. The large y-axis intercept may be correlated with a possible structural phase transition of the Ho samples at 8% Ho content (cp. Appendix A). However, a detailed investigation of the sample structure is missing.

A increase of the damping parameter α with increasing amount of Dy can be seen in NiFeDy samples (cp. Fig. 6.1 a)). The y-axis intercept again is small for 2.5% Dy and higher for 5% Dy, which again is a indicator that the intrinsic damping dominates for smaller RE contents. Note that it is not possible to obtain reasonable information for low frequencies of samples with contents of >10% Ho or >5% Dy, as the linewidth increases drastically and the amplitude of the absorption signal becomes smaller as the amplitude is inverse proportional to the linewidth in VNA-FMR. This is also the explanation why the number of experimental points decreases for frequencies in the range up to 10 GHz for higher Ho/Dy concentrations as only a few data sets could be evaluated reasonably.

In Fig. 6.1 b) the results of the NiFeGd samples are plotted. An increase of the damping is obvious, but compared to Ho and Dy it is much smaller (note the different y-scales). The nearly zero y-axis intercepts are indicators for the vanishing extrinsic damping contribution and the good sample quality.

All values for the damping parameter α as obtained by in-plane FMR at a frequency of 21.9 GHz are summarized in Fig. 6.2.

The influence of Gd on the damping is much less than the one of Ho and Dy. A linear dependence of the damping parameter α on the RE content is found for the series of NiFeGd and NiFeDy samples. The damping parameter of the NiFeHo series also depends linear on the Ho content but the dependence is different for the interval of 0%-8% and 8%-16% Ho content. This may be correlated with the possible structural phase transition of the Ho samples at 8% Ho content (cp. Appendix A). As expected from prior results by Reidy et al. [11], Gd does not change the value of α drastically. This can be understood using Hund's rules and the fact that the orbital angular moment L is zero for Gd (cp. chapter 1). For Gd the damping channel due to spin-orbit coupling is eliminated.

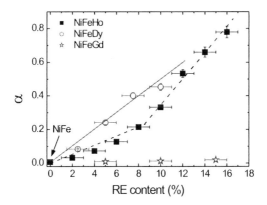

Figure 6.2: Results of in-plane FMR measurements at 21.9 GHz for different NiFe, NiFeGd, NiFeHo and NiFeDy samples. A linear dependence of the damping parameter α on the RE content for the NiFeGd and NiFeDy series is found. A linear dependence is also found for the NiFeHo series but the dependence is different in the region of 0%-8% and 8%-16% Ho content (note the different slopes of the dashed line).

The increase in α for the Gd samples is mainly caused by the reduction of M_s as $\alpha = 1/(\lambda\mu_0\gamma M_s)$ (cp. subsection 2.3.1). γ increases slightly as g increases from 2.12 for NiFe up to g=2.27 for NiFeGd15, which is more than compensated by the drop of M_s from 807 kA/m (807 emu/cm^3) for NiFe to 212 kA/m (212 emu/cm^3) for NiFeGd15 (cp. Appendix A). λ can be assumed to be constant as it is based on the intrinsic damping mechanism that should not change drastically.

Due to Hund's rules the largest increase of the damping parameter α is expected for Ho, as the total angular momentum J and orbital momentum L both have their maximum value. Theoretical investigations by means of a spin polarized relativistic Korringa-Kohn-Rostoker (SPR-KKR) model by H. Ebert on NiFeRE samples emphasize the validity of Hund's rules that are derived for isolated RE ions [130]. This was expected as the 4f-shells are far below E_F.

However, in our measurements we find the largest damping increase for Dy doping. At least for the small doping regime (< 5% RE content) extrinsic damping can not be the origin of the increased damping as the y-axis intercepts in Fig. 6.1, and thus the extrinsic damping, are nearly zero. Hence, the interesting and unexpected result is that the a certain amount of Dy enlarges the magnetic damping more than the same amount of Ho. A possible explanation for this behavior could be different sample structures (clusters, inhomogeneities etc.). By means of depth profile analysis we find that increasing dopant concentrations lead to increasing inhomogeneities of the magnetic films. However, this is no explanation for the unexpected behavior of the samples with RE contents < 5%.

Nevertheless, the measurements show that it is possible to modify the dynamic magnetic behavior, especially the damping, of NiFe samples by different RE elements and by the amount of doping. In contrast to the findings of Reidy et al. [11] we see, that not only Hund's rules have to be taken into account. Also the exact knowledge of the sample properties and the sample structure seems to be crucial.

To obtain more insight in the intrinsic and extrinsic contributions to the damping parameter α, FMR experiments in the perpendicular configuration were carried out. In this configuration two magnon scattering is switched off [32] (cp. subsection 4.1.1). However, it is harder to measure in the perpendicular configuration, in which the sample is placed at the side of the wave guide because the excitation field amplitude is smaller (e. g. due to different field distributions). More details on this topic are given in [130]. Hence, it was not possible to obtain reasonable results in the perpendicular configuration for samples with RE contents of > 6% Ho or > 2.5% Dy. The results of the perpendicular FMR experiments at a frequency of 21.9 GHz are summarized in Table 6.1. In addition the results of the in-plane FMR measurements that are shown in Fig. 6.1, are listed to simplify the comparison. In the case of NiFe and NiFeGd the value of the in-plane- and out-of-plane α are equal within the accuracy of the measurements. As indicated by the measurements presented in Fig. 6.1 the intrinsic contribution dominates. The same behavior is found for the NiFeHo samples up to 6% Ho content, except NiFeHo2. Therefore the damping for these samples is predominantly intrinsic which is in agreement with the findings of Fig. 6.1. For higher Ho/Dy contents no feasible results for out-of-plane FMR are available but according to Fig. 6.1 it is very likely that the extrinsic contributions rise as the y-axis intercepts are larger. This may be attributed to inhomogeneities of the NiFeRE films for high doping which we found in the structural analysis.

sample	α in-plane (FMR)	α out-of-plane (FMR)	α (TR-MOKE)
NiFe	0.0076 ± 0.0004	0.0069 ± 0.0004	0.018 ± 0.003
NiFeGd5	0.0088 ± 0.0005	0.0089 ± 0.0005	0.029 ± 0.004
NiFeGd10	0.0117 ± 0.0006	0.0118 ± 0.0006	0.033 ± 0.004
NiFeGd15	0.0187 ± 0.0009	0.0189 ± 0.0009	0.038 ± 0.005
NiFeHo2	0.038 ± 0.002	0.0270 ± 0.002	0.047 ± 0.005
NiFeHo4	0.070 ± 0.004	0.063 ± 0.004	0.060 ± 0.009
NiFeHo6	0.127 ± 0.006	0.120 ± 0.006	—
NiFeHo8	0.21 ± 0.01	—	—
NiFeHo10	0.33 ± 0.01	—	—
NiFeHo12	0.53 ± 0.03	—	—
NiFeHo14	0.66 ± 0.03	—	—
NiFeHo16	0.78 ± 0.04	—	—
NiFeDy2.5	0.083 ± 0.005	0.077 ± 0.005	0.081 ± 0.09
NiFeDy5	0.24 ± 0.01	0.24 ± 0.01	—
NiFeDy7.5	0.40 ± 0.02	—	—
NiFeDy10	0.45 ± 0.02	—	—

Table 6.1: Damping parameter α as extracted from the in-plane and perpendicular FMR configuration, and from the TR-MOKE experiments. All FMR measurements are performed at a frequency of 21.9 GHz.

For more information about the magnetization dynamics TR-MOKE experiments were performed. Fig. 6.3 summarizes the results on the NiFeGd, NiFeHo and NiFeDy series. The results for the damping parameter α are also listed in Tab. 6.1. The data analysis was performed as described in chapter 5. All measurements were performed on the magnetic sample holder to prepare the canted magnetic state which is required for the excitation. Note that the angle of the magnetization θ_M (cp. Fig. 5.2) is not constant as in chapter 5 but depends on the investigated sample as the saturation magnetization M_s changes with the amount of doping (cp. Appendix A). θ_M is maximal for the NiFeGd15 sample with $\theta_M = 31°$. However, numerical simulations show that θ_M does not significantly influence the results for the damping parameter α obtained via equation

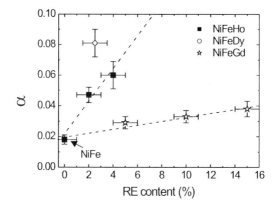

Figure 6.3: Effective damping factor α for different RE dopants as function of RE content. A linear increase of α with respect to the RE content is found. The lines are guides to the eye.

(4.18) which is in agreement with findings by Djordjević [128]. A linear dependence of the damping parameter α on the RE content is found for the Gd and Ho doped sample series. Gd doping only slightly changes the magnetic damping. As it was not possible to obtain reasonable TR-MOKE data for a doping of $> 4\%$ Ho and $> 2.5\%$ Dy no statement for the Dy series can be given. Nevertheless, the higher damping of Dy doped samples compared to Ho doped samples as found by FMR is supported by the TR-MOKE experiments. The damping parameter α in the TR-MOKE is evaluated to be 2-3 times larger for the Gd doped samples than the values of the damping parameter in FMR. The reasons for that behavior are the same as explained in chapter 5. However, for the Ho and Dy series a good agreement of the damping parameters α as evaluated by TR-MOKE and FMR exist. This is due to the fact that with increasing α the decay time τ in the TR-MOKE measurements decreases and therefore the influence of the beam movement due to the mechanical delay stage becomes negligible.

6.2 Conclusion

(VNA-) FMR and TR-MOKE techniques were used to analyze the magnetization dynamics of RE-doped NiFe samples. The influence of Gd on the damping is found to be minimal as expected due to considerations on Hund's rules. Ho and Dy lead to large changes of the value for the damping parameter α, even for small RE contents. Surprisingly, Dy influences the magnetic damping more than Ho. Possible reasons for this behavior at higher doping levels ($> 5\%$) can be impurities and structural defects in the samples as first structural analysis show large inhomogeneities in highly doped samples. However, the main contribution to damping is intrinsic in the small doping regime ($< 5\%$), merely small extrinsic contributions are present. Thus extrinsic damping can not explain why the additional damping contributed by Dy is higher than the one by Ho.

Beside the higher values for the damping parameter α in the Gd doped series, the results of TR-MOKE and (VNA-) FMR are in good agreement. In all measurements we find a linear dependence of the damping parameter α on the RE content. For the Ho doped samples the dependence is different on the interval of 0%-8% and 8%-16% Ho content. A structural phase transition may be the reason for this.

Considering all results we do not have an explanation for the finding that Dy doping leads to a larger damping than Ho doping. Two magnon scattering can be excluded to be the reason. Maybe the outstanding detailed investigation of the structural properties of the samples can reveal the mechanism. Altogether, RE dopants are a reasonable way to influence the magnetization dynamics of NiFe.

7 Experiments on CoGd

In this chapter we will focus on a special case of a ferrimagnet, namely CoGd. As Gd has a g-factor of 2 [131], and therefore essentially zero orbital momentum, the contribution to damping via the spin-orbit coupling is negligible [11]. Thus it is possible to study the effective damping parameter α_{eff} as a function of sample composition and temperature. As introduced in section 2.4, there are an angular momentum compensation and a magnetic compensation point. In CoGd these points are close together, which is another reason for the choice of that material system. The effective damping α_{eff} and the effective gyromagnetic ratio γ_{eff} are expected to increase in the compensation region. The results of magnetization dynamics measurements of frequency and time domain methods around the compensation points are presented and compared.

Three sets of amorphous samples are used. In the following sample set (A) denotes a CoGd gradient sample on Si and sample set (B) a CoGd gradient sample on GaAs. The gradient samples were cleaved into small pieces for the measurements. They are used to examine the static and dynamic magnetization properties mainly as function of composition. The third set of sample (C) consists of specimens with fixed concentrations, which allow to mainly study the magnetization as a function of temperature.

7.1 Static Magnetic Measurements

First, we want to concentrate on the static magnetic properties of the samples. Figure 7.1 contains information about the magnetic moment normalized to the magnetic moment at zero Kelvin as a function of temperature and Co concentration. The data were measured by superconducting quantum interference device magnetometery (SQUID). Obviously, the temperature of the magnetization compensation point T_{M} shifts to lower temperatures with increasing Co content.

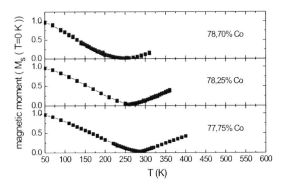

Figure 7.1: Remanent magnetization as measured by SQUID as a function of CoGd alloy composition and temperature. All data shown in this figure were obtained from sample set (B).

For sample set (B) we observe T_M at room temperature at a Co concentration of about 77.5 %. This is in agreement with prior published data on CoGd [90]. The results show that we are able to tune T_M by means of the composition of the samples. In sample sets (A) and (C) the room temperature magnetization compensation occurs at slightly different concentrations. Also the value of the coercive field varies for different samples. This can be explained by (small) variations of the growth conditions and sample properties.

In Fig. 7.2 a) the coercive field H_c as a function of temperature and Co concentration is plotted. The data are obtained by static MOKE. At a fixed Co concentration we find an increase of H_c at T_M. This is expected as the net-magnetization of the ferrimagnet tends to zero and therefore a higher field is necessary to exert a torque on the magnetization, and to switch the magnetic state (cp. section 2.4).

In Fig. 7.2 b) a hysteresis loop below (left loop) and above (right loop) the magnetization compensation point is shown. The hysteresis loops change sign as one passes T_M from one side of the magnetization compensation point to the other. There are two reasons for that behavior.

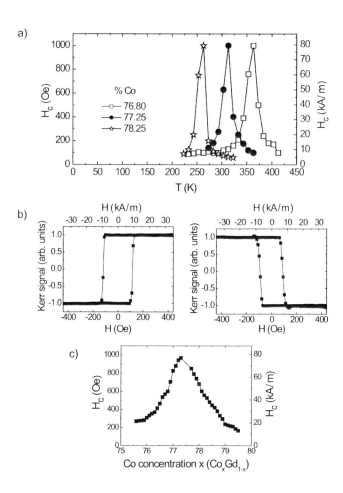

Figure 7.2: Magneto-optic Kerr effect measurements. a) shows the coercive field H_c as determined as a function of CoGd-alloy composition and temperature. The left (right) hysteresis loop in b) is measured below (above) the magnetization compensation point. Note the reversal of the hysteresis loops. In c) the coercive field H_c vs. the Co concentration at RT is depicted. All data are obtained from sample set (B), except the ones of the lower panel, which are from sample set (A).

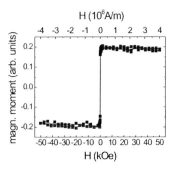

Figure 7.3: High-field hysteresis loop as measured by SQUID at RT for a piece of sample set (A). The magnitude of the switching field is smaller 119 kA/m (1.5 kOe). No further switching or change of the magnetic moment is found for larger magnetic fields, which shows that the exchange coupling well exceeds $4 \cdot 10^6$ A/m (50 kOe).

First, the laser in all MOKE and TR-MOKE experiments predominantly probes and excites the Co sublattice. This important feature will be explained in detail in the next section. And second, the antiferromagnetic exchange coupling is much larger than the externally applied magnetic fields. In the case of CoGd the exchange coupling well exceeds $4 \cdot 10^6$ A/m (50 kOe) and therefore all accessible magnetic fields in our setup. This was verified by high field hysteresis loops that were measured by SQUID. In Fig. 7.3 a high field hysteresis loop is plotted. The magnitude of the switching field is smaller 119 kA/m (1.5 kOe). No further switching or change of the magnetic moment is found in larger applied magnetic fields. This shows that the exchange coupling exceeds the the maximum accessible magnetic field of the SQUID, which is $4 \cdot 10^6$ A/m (50 kOe).

Hence it is easy to understand why the hysteresis loops change sign at the magnetization compensation point. Below T_M the Gd sublattice is dominant and therefore aligned parallel to the external field. Above T_M the Co sublattice is dominant and aligned with the applied magnetic field. As the laser predominantly detects the direction of the Co sublattice the sign of the loops changes.

In Fig. 7.2 c) the coercive field H_c versus the concentration at room temperature for sample set (A) is plotted. A clear increase of the coercive field H_c is found. The position of the peak of about 77.25% Co is in agreement with the

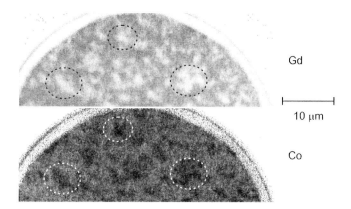

Figure 7.4: PEEM images of a homogenous CoGd alloy with $T_M = 350$ K. The contrast, brightness and intensity of the images was changed to enhance the contrast of the domains. Black/white indicates orientation of the magnetization in opposite directions. Some corresponding domains of the two sublattices are marked for clarity. The antiferromagnetic coupling of the domains can be recognized.

magnetization compensation that was determined using SQUID.

During the thesis is was possible to perform some X-ray Magnetic Circular Dichroism (XMCD) [132] and Photo Emission Electron Microscopy (PEEM) [133, 134] measurements on CoGd samples at the surface/interface microscopy (SIM) beam line of the Swiss Light Source (SLS) [135]. For element specific detection of Co the energy of the X-rays was tuned to 786 eV (L-edge) [136] and for Gd to 1197 eV (M-edge) [137]. Only static data could be obtained and evaluated. Fig. 7.4 shows a PEEM measurement at RT on a homogenous 50 nm thick CoGd sample on a glass substrate, covered with 1.5 nm Ta and 2 nm Pt. T_M is 350 K. It turned out that the cap layer material and thickness only allow to measure small XMCD signals. However, some basic results can be derived. The PEEM measurements reveal irregular domains with diameters of few µm. In the domains a reversed contrast between the Co and Gd signal is found, which emphasizes the antiferromagnetic coupling of the Co and Gd magnetic moments.

Furthermore, XMCD measurements were carried out as a function of the applied

magnetic field and the temperature. The result are hysteresis loops at different
temperatures which reveal the switching behavior of the Co- and Gd-sublattice,
respectively (cp. Fig. 7.5). The probed area of the sample was again heated

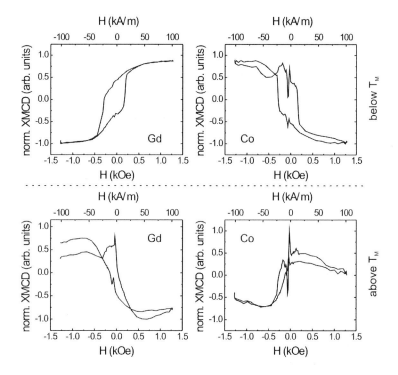

Figure 7.5: Hysteresis loops of a homogenous CoGd alloy with $T_M = 350$ K below
and above T_M. The XMCD signal is normalized. Note that the orientation of Co and
Gd loops is always opposite. At T_M the orientation of the loops changes.

by the cw-laser. With increasing temperature the coercive field H_c increases,
and at T_M the switching of the hysteresis loops is observed as expected. The
sense of corresponding Co and Gd hysteresis loops below and above T_M are always opposite which again is based on the antiferromagnetic coupling of the
sublattices.

7.2 Co-Selectivity of the MOKE Setup

In the last section we argued that the laser beams in our optical experiments predominantly excite and probe the Co sublattice. This is based on the arguments that will be presented in detail in the following. In CoGd alloys the Co spin-split bands are close to the Fermi-level (E_F). By contrast, the spin-split bands in Gd are spatially localized and lie $\sim$8 eV below E_F [138,139]. In the CoGd ferrimagnet the Co-moment is mostly carried by the spin-split 3d band near E_F. The effective moment of Gd close to E_F comprises less than 5% of Gd's total moment which is principally contained in the deep spin-split 4f states [138]. In our experiments we use laser radiation within a wave length region of 420 nm (2.95 eV) to 840 nm (1.48 eV). Therefore one mainly detects the magnetic signal of the Co sublattice. In the case of pure optical excitation in TR-MOKE experiments, the laser pulse rapidly disturbs electrons near E_F in the sample. After thermalization, energy is subsequently transferred to the spin and phonon systems leading to a sudden change of the magnitude and/or direction of the internal magnetic field $\boldsymbol{H_{eff}}$ following the mechanisms described by the 3-T model (cp. subsection 2.5.2). Apart from ultra fast demagnetization, the modified internal field can cause the magnetization to precess around a new equilibrium position thereby triggering the observed precessional motion of the magnetization as described in subsection 2.5.3. In the case of CoGd the laser pulse energy is initially deposited into the Co sublattice near E_F. When the Co sublattice – which is AF exchange coupled to the Gd subsystem by an exchange field well in excess of $4 \cdot 10^6$ A/m (50 kOe) – is excited, the Gd follows its precessional motion 180° out of phase. In the specific case when the system is fully compensated, the total saturation magnetization equals zero (M=0) irrespective of whether the moments are moving or not. Thus, with a measurement that is sensitive to the total magnetization $\boldsymbol{M}$ it is difficult, if not impossible, to investigate the dynamics of the independent sub-lattices near the compensation points. By contrast, in the laser pump-probe experiment, both the excitation and the probe pulses couple predominantly to the transition metal and consequently magnetization oscillations of the two coupled sublattices at the magnetization compensation point does not violate the mean-field M = 0 constraint. Thus, the pump-probe technique may reveal the magnetization dynamics of a system with zero total local magnetic moment. However, due to the inhomogeneous heating of the laser (laterally as well as across the film thickness) and the gradient of some samples, the magnetization in the sample will never be completely compensated. Hence a 'real' measurement at the magnetic compensation point is not possible, which results in a smearing out of the data. Analogously it is not possible to perform a 'real' measurement at the angular compensation point.

7.3 Magnetization Dynamics

After the static characterization of the samples, the results of the dynamic measurements of the CoGd ferrimagnet will be presented now. According to the LLG-equation for a strongly coupled ferrimagnet (2.29) in section 2.4, we expect the effective damping parameter α_{eff} and the precession frequency f of the system to diverge at the angular compensation point. Temperature dependent FMR, as well as temperature and composition dependent TR-MOKE measurements were carried out to verify these predictions. Furthermore VNA-FMR and BLS experiments were performed. In the dynamic measurements we focus on the precessional motion and its damping behavior on ps timescales.

7.3.1 VNA-FMR and FMR Results

First the results of the FMR methods are discussed as these methods excite and probe the net-magnetization of the samples. Sample set (B) was used as it provides a smaller concentration gradient. Note that it is not possible to get reasonable data of the homogenous sample set (C) as the sample thicknesses are to small to provide sufficient magnetic signal, particularly in the vicinity of the compensation points, where the magnetization is very small.

The dynamic properties in the frequency domain were characterized with FMR at 22 GHz between 250 and 380 K. The VNA-FMR is only probing a 90 µm wide stripe of the sample at frequencies between 45 MHz - 20 GHz. Figure 7.6 summarizes the frequency domain results of both FMR setups. We plot the effective g-factor extracted from the VNA-FMR frequency measured in an external field of 71.6 kA/m (900 Oe) as a function of Co concentration (75.5% to 80.0%) in Fig. 7.6 a) and observe the characteristic increase of the measured resonance frequency (or g-factor) for M at a Co concentration of 78%. Instead of f the effective g-factor is plotted, which is calculated using the well known relations

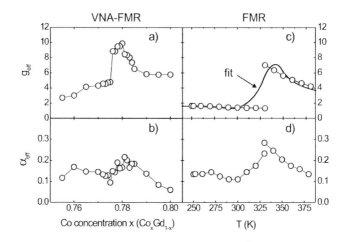

Figure 7.6: Left: a) and b) Data obtained from VNA-FMR measurements as a function of CoGd alloy composition. a) effective g-factor g_{eff}, b) effective damping parameter α_{eff}. Measurements were performed at room temperature and in an applied field of 71.6 kA/m (0.9 kOe) for VNA-FMR. Right: Data extracted from temperature dependent FMR measurements at 22 GHz. c) effective g-factor g_{eff} versus temperature and d) effective damping parameter α_{eff} versus temperature. The solid line in c) corresponds to a fit of g_{eff} using γ_{eff} given in eqn. (2.29). All data shown were obtained from sample set (B). The error bars ($\Delta g_{eff}/g_{eff} = \Delta\alpha/\alpha = 5\%$) are omitted for clarity.

$$\gamma_{eff} = g_{eff}|e|/2m_e, \tag{7.1}$$

$$\omega = \gamma_{eff}\mu_0 H_{eff}. \tag{7.2}$$

The use of g_{eff} instead of the frequency f eases the comparison of the results of different measurements, as the g-factor does not depend on the magnitude of the applied magnetic field as f. In Fig. 7.6 b) we plot the measured α_{eff}, which is calculated as $\alpha_{eff} = \Delta f/f$. This relation can be derived from equations (4.11) and (2.17), and is valid, if the magnitude of $\boldsymbol{M}$ can be neglected. This is the case near the magnetization compensation point. Fig. 7.6 c) and d) plots g_{eff} and α_{eff} as function of temperature from FMR measurements. For the determination of α_{eff} equation (4.3) is used. In both cases a significant increase of g_{eff} and α_{eff} around the angular compensation point is observed. The singularity, i. e. the divergence in both quantities according to equation (2.29), is bounded in real samples due to sample inhomogeneity (RE concentration) and temperature gradients as the temperature approaches T_L. This means that the sample is never completely compensated, which also explains why the frequency f does not drop to zero as predicted by the formula. The values for g_{eff}, which were obtained by FMR, where fitted according to equation (2.29). The fit results are indicated in Fig. 7.6 c) by the solid line. Based on the SQUID measurements (cp. Fig. 7.1 a)) we chose a linear decreasing magnetization for both sublattices with increasing temperature T for the fit. The saturation magnetization at $T = 0$ K and the g-factor for the fit are $M_s^{Co} = 1400$ kA/m (1400 emu/cm^3) [21] and $g_{Co} = 2.18$ [140] for Co and $M_s^{Gd} = 2060$ kA/m (2060 emu/cm^3) [21] and $g_{Gd} = 2$ [131] for Gd, respectively. Due to the concentration gradient present in the sample, the magnetization can never be completely compensated in the FMR measurements. For sample (B) the concentration gradient leads to a spread of RE concentration of about 0.2% and corresponds to a spread of T_L of about 20 K according to Hansen et al. [90]. This effect was taken into account for the fit. We find a separation of T_M and T_L of about 30 K. As can be seen in Fig. 7.6 c) the fit reproduces the FMR measurements reasonably well. Assuming the same parameters and a typical value of $1 \cdot 10^4$ for λ_{ex} [58], the equation for the exchange mode of the ferrimagnetic system (cp. eqn. (2.31)) predicts the optical exchange mode (cp. Fig. 2.10) with a minimum frequency of 980 GHz around the angular momentum compensation temperature T_L. Therefore it is not possible to detect this mode in our setups, if it is exited at all.

7.3.2 BLS and TR-MOKE Results

Now we want to discuss the results of the optical experiments. In the laser pump-probe TR-MOKE experiment we use the 840 nm (1.48 eV) laser pulses with 170 fs duration at 260 kHz repetition rate to trigger the precession of the magnetization in an externally applied magnetic field. For the measurement the magnetic sample holder was used. Altogether a static magnetic field of 175 kA/m (2200 Oe) was applied under an angle of 60° with respect to the sample plane (cp. Fig. 7.7). The heating pulse is focused into a 50 µm diameter spot and a

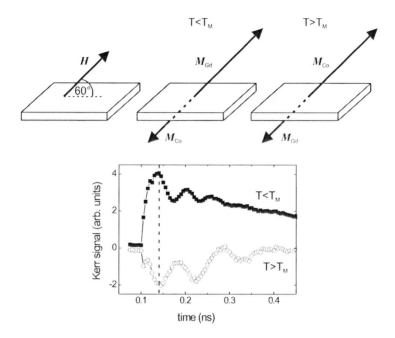

Figure 7.7: Alignment of the sublattice magnetizations of CoGd above and below T_M: the external magnetic field $\boldsymbol{H}$ is applied under an angle of 60° with respect to the sample plane. For $T < T_M$ the magnetization of the Gd-sublattice $\boldsymbol{M}_{Gd}$ is dominant and therefore oriented along the direction of the applied field. For $T > T_M$ the Co-sublattice is dominant and $\boldsymbol{M}_{Co}$ is aligned along the direction of the applied field. This leads to a change of sign and a phase shift of 180° for the Kerr signals of sample (C) measured above and below T_M (see dotted line), as predominantly the out-of-plane component of the Co-sublattice is probed.

pump fluence of 15 mJ/cm² is used. The precession of the magnetization is monitored using the polar Kerr effect with a frequency doubled laser pulse at 420 nm (2.95 eV) focused into a beam diameter of 15 µm resulting in a low fluence of < 1.5 mJ/cm². As explained in subsection 7.2 the laser pulses predominantly excite and probe the Co-sublattice. To further investigate the magnetization dynamics in the vicinity of the compensation points we now focus on a sample

(C) with constant composition of 78% Co. From VSM measurements we expect T_M to be around 390 K in this sample.

The alignment of the sublattice magnetizations M_{Gd} and M_{Co} of CoGd and an example of a TR-MOKE measurement above and below the magnetization compensation point T_M are shown respectively in Fig. 7.7. For $T < T_M$ the magnetization of Co is smaller than that of Gd. Thus M_{Gd} is aligned along the direction of the applied field. Above T_M the magnetization of the Co-sublattice is larger than that of the Gd-sublattice and the ferrimagnetic system flips, allowing M_{Co} to be aligned along the direction of the applied field H. In our experiment predominantly the out-of-plane component of the Co-sublattice is probed. This leads to the observed sign change of the Kerr signal and a phase shift in the precession of 180° when crossing the magnetic compensation point from low temperature.

By the time-resolved pump-probe experiments T_M is confirmed to be around 390 K, see the large frequency increase in the vicinity of the magnetization compensation temperature in Fig. 7.8 a).

The decay time τ decreases by about 50% (cp. Fig. 7.8 b). Note, that the 'static' temperature increase by the laser heating (approximately 50 K) is already included in the temperature presented here. The temperature increase can be approximated by measuring hysteresis loops with and without pumping by the red beam. The coercive field H_c with and without pumping can be attributed to a temperature, if data for H_c as function of the temperature exist (cp. Fig. 7.2 a)). The increase in α_{eff}, which corresponds to a decrease in τ, is shown in Fig. 7.8 d). α_{eff} is evaluated using equation (4.18). This relation is valid in the vicinity of the compensation points as the magnetization is small.

The peak of of the damping parameter at the angular momentum compensation point is more pronounced in the data obtained using sample set (A), see Fig. 7.9 a). TR-MOKE was used to measure α_{eff} as a function of the Co concentration at room temperature on the the non-magnetic sample holder. Equation (4.18) is used for the evaluation of α_{eff}. In addition the behavior of the damping parameter as a function of the composition as measured by BLS is given for sample set (B) in Fig. 7.9 b). Again an increase of α_{eff} in the region of 78% Co is found as expected. α_{eff} is extracted from the BLS data via $\alpha_{eff} = \Delta f / f$. Note that this is the same relation as in the case of VNA-FMR. The relation can be

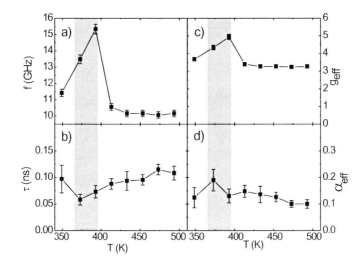

Figure 7.8: Summary of the time resolved pump-probe data. In a) and b) we report the measured frequency and decay time of the oscillations of the magnetization as a function of temperature for sample (C) ($Co_{78}Gd_{22}$). The measurements are carried out in a field of 175 kA/m (2200 Oe) applied at 60° with respect to the sample plane. The data is extracted from traces similar to those shown in Fig. 7.7. The error bars are obtained by the scatter of the fit results when using different methods to remove the background. The grey colored areas indicate the position of the compensation region as determined from the divergence of the coercive field in static hysteresis loops and uncertainties in the temperature of the sample. c) and d) show the corresponding effective g-factor g_{eff} and damping constant α_{eff}.

used as BLS and VNA-FMR both are frequency domain techniques (cp. section 4.2). The precessional frequency in both measurements also shows an increase in the compensation region.

The gradient of sample set (A) and (B) and the inhomogeneous heating by the laser (laterally as well as across the film thickness) lead to a smearing out of the data in the TR-MOKE and BLS experiments. The singularity, i. e. the divergence of α_{eff} according to equation (2.29), is bounded due to these inho-

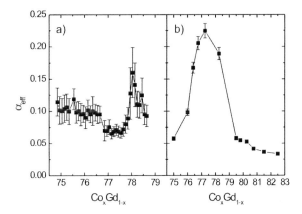

Figure 7.9: Left: Dependence of α_{eff} versus Co concentration for sample set (A) as measured by TR-MOKE. Right: Dependence of α_{eff} versus Co concentration for sample set (B) as measured by BLS.

mogeneities, and the sample is never completely compensated in TR-MOKE experiments.

In conclusion the increasing precession frequency and the increasing damping constant in the vicinity of the compensation points are consistent with the mean field model developed in subsection 2.4. Regardless how the magnetic film was excited, by an external magnetic field (FMR) or an optical laser pulse (TR-MOKE), similar results are obtained. The temperature dependence of the TR-MOKE results by Stanciu et al. [141] agree with our findings. We also find the increase of the precessional frequency f and of the damping parameter α_{eff} at the compensation point T_{L}.

However, in contrast to their results we do not observe any evidence of a high frequency optical exchange mode. Far from the compensation region the optical exchange mode is in the range of several THz. The mean field model equation (2.31) predicts that the frequency of the exchange mode drops to zero at the angular momentum compensation point. As a consequence of temperature and concentration gradients (laterally as well as across the film thickness) the divergence of the resonance frequency around the angular momentum compensation point is always limited (cp. Figs. 7.6 and 7.8). Based on the mean field model which leads to the fit shown in Fig. 7.6 c) we expect that the exchange mode frequency does not drop below several hundred GHz at T_L in our samples. It seems unlikely that in pump-probe TR-MOKE experiments the optical mode can be excited coherently. This is in agreement with the fact that we never observed the exchange mode. It may be possible to detect this mode in BLS since there the thermal excitation is measured and coherence is not required. However, in our BLS data no signature of a high frequency mode can be found.

7.4 Conclusion

We have analyzed the damping and frequency behavior of a model ferrimagnetic system, namely CoGd, in the vicinity of the magnetization and angular momentum compensation point. We have shown that by using complementary techniques it is possible to study the dynamic behavior of the total magnetization in detail. We observe a large increase in precessional frequency and damping for the total magnetization in the frequency and the time domain. This is in agreement with the mean-field model, that was introduced in section 2.4. CoGd allows one to tune the precessional frequency f and the damping parameter α_{eff} due to the sample composition and the temperature of the sample. The time domain study shows that an all-optical control of the magnetization dynamics in the vicinity of the compensation points may be feasible. CoGd, and in general ferrimagnets close to the compensation points are therefore well suited for high speed manipulation of the magnetization and may be of interest to magneto-optic recording applications.

8 Experiments on FeGd

In this chapter the result of TR-MOKE and FMR measurements of a FeGd multilayer sample are discussed. The measurements have been inspired by previously performed X-PEEM measurements at the Advanced Light Source (ALS) in Berkeley. Surprisingly, the X-PEEM measurements showed decay times of several nanoseconds for the magnetization using very small pump fluences (< 1 mJ/cm^2). The aim was to verify the results of the X-PEEM experiment by TR-MOKE and FMR using exactly the same sample.

There are further interesting aspects for the investigation of FeGd. One point is that FeGd is also a RE/TM-ferrimagnet with negligible SO-coupling due to the zero orbital moment of Gd [11, 131]. Furthermore, FeGd multilayers can have a Spin Reorientation Transition (SRT). A SRT in a thin ferromagnetic film is a change in the easy axis of the magnetization induced by varying the temperature, film thickness or composition. This can be explained by changes in the demagnetizing energy and by the temperature dependence of both, the surface and the volume anisotropy. The ratio of these anisotropies can be tailored in magnetic multilayers by varying the thickness of the individual layers. Hence, for FeGd multilayers the easy axis of the ferromagnetic film can change from in-plane to out-of-plane or vice versa in a certain temperature range [142, 143]. The canted magnetic state that exists around the SRT is well suited to trigger magnetization dynamics in all-optical pump-probe experiments (cp. subsection 2.5.3). Ultrafast switching of the direction of the magnetization, e. g. from in-plane to out-of-plane may be possible in all-optical pump-probe experiments.

8.1 Sample Preparation and Properties

The FeGd sample was grown by rf-ion beam sputtering from Fe and Gd targets on a glass substrate at an Ar pressure of $1 \cdot 10^{-4}$ mbar. T he layer structure is given by Al(0.8nm)/[Gd(0.5nm)/Fe(0.5nm)]$_{12}$Al(0.8 nm). The Al cap layer prevents oxidation and is thin enough to yield a sufficient photoemission signal in the X-PEEM experiments.

Measurements on a similar sample than the one we investigated show a SRT around 320 K. However, our sample was characterized using temperature dependent SQUID after the dynamical experiments and no signature of a SRT was found. In our case a canted magnetic state exists, i. e. the magnetization has a non-zero in-plane and out-of-plane component. With increasing temperature both, the in- and out-of-plane component of the magnetization get smaller as expected for an usual ferromagnetic material. The missing SRT is very likely due to the inhomogeneity of the sample. Also in the X-PEEM experiments a SRT was found only at certain sample positions. As X-PEEM is a very local method (the lateral resolution is about 50-100 nm) a sample position showing a SRT can be found. Furthermore, TR-MOKE experiments as well as SQUID measurements have been performed more than a year after the X-PEEM measurements. Hence aging of the sample can not be excluded.

Like the CoGd-alloys in chapter 7, FeGd-multilayers also have a magnetization compensation point [144]. For our sample we do not observe the signature of a narrow compensation region. This is also very likely correlated to the inhomogeneities in the sample. At room temperature the magnetization is around 100 kA/m (100 emu/cm^3) and decreases to about 70 kA/m (70 emu/cm^3) at 350 K. The magnetization lies mainly in-plane.

The pump fluence for the pump-probe experiments was usually about 5 mJ/cm^2. From experiments on the CoGd samples we can estimate the average heating of the sample to about 20 K, i. e. in the pump-probe experiments presented in the following section M_s should be around 90 kA/m (90 emu/cm^3). All TR-MOKE experiments presented in this chapter were carried out using the non-magnetic sample holder since the small out-of-plane component allows a sufficient excitation in the TR-MOKE experiments.

In-plane hysteresis loops measured by MOKE reveal that the coercive field H_c is depending on the sample position and varies in a range of approximately 3.98 - 7.16 kA/m (50-90 Oe). Fig. 8.1 shows an easy axis in-plane hysteresis loop with $H_c = 5.2$ kA/m (65 Oe). Furthermore, typical domains with diameters of some µm up to 150 µm were found in the PEEM observations [145]. Thus most TR-MOKE experiments were performed in a sufficiently large applied magnetic field (> 7.9 kA/m (100 Oe)) to make sure that the sample is in a single domain

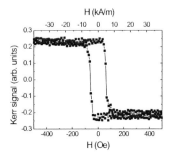

Figure 8.1: In-plane hysteresis loop as measured by MOKE with $H_c = 5.2$ kA/m (65 Oe).

state. In this case effects due to different domain configurations can be excluded.

8.2 Magnetization Dynamics

Now we want to focus on the magnetization dynamics of the FeGd multilayer. In Fig. 8.2 a) TR-MOKE measurements for two different applied in-plane magnetic fields are shown. As expected a clear increase of the precession frequency is observed with increasing magnitude of the applied magnetic field. Further, the smaller background compared to measurements in the previous chapters is noticeable. The background is found to depend strongly on the strength and the direction of the magnetic field and the sample position. As introduced in chapter 5, equation (5.2), the decay time τ_b of the exponential background is determined using the relation

$$y(t) = A \cdot \exp(t/\tau_b) + y_0. \tag{8.1}$$

The fit according to equation (8.1) for the trace with the in-plane field of 79.6 kA/m (1000 Oe) yields $\tau_b = 350 \pm 140$ ps and is also plotted in Fig. 8.2 a). Further TR-MOKE measurements as a function of the applied magnetic in-plane field are presented in Fig.8.2 b).

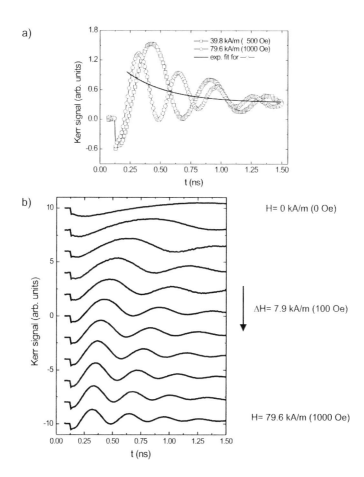

Figure 8.2: a) Kerr signal for an applied in-plane magnetic field of 39.8 kA/m (500 Oe) and 79.6 kA/m (1000 Oe) with a precessional frequency f=1.9 GHz and f=3.1 GHz, respectively. A fit to 79.6 kA/m (1000 Oe) trace according to equation (8.1) with $\tau_b = 350$ ps is shown. b) Kerr signal as a function of the applied in-plane magnetic field. The traces are offset for clarity. The magnetic field is successively increased in steps of ΔH=7.9 kA/m (100 Oe) from 0 kA/m (0 Oe) (highest trace) to 79.6 kA/m (1000 Oe) (lowest trace).

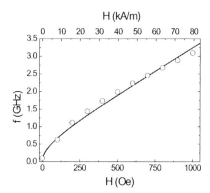

Figure 8.3: Frequency versus applied in-plane magnetic field. The open circles represent the measured frequency f, the solid line corresponds to a fit using equation (2.19) with g=1.5 and $M_s = 98$ kA/m (98 emu/cm^3).

Note that only data that are obtained at the same sample position can be directly compared to each other. This is due to the sample inhomogeneity and the corresponding differences in the measured TR-MOKE signals. Consequently, the obtained data only characterize a certain sample position. This also means, that discrepancies to the X-PEEM experiments may be attributed to different sample properties. All TR-MOKE results presented in this chapter are measured at the same sample position.

In Fig. 8.3 a) the precession frequency (open circles) versus the applied in-plane magnetic field and the corresponding Kittel fit (solid line) according to equation (2.19) are plotted. The fit with the free parameters M_s and g yields g=1.5±0.05 and $M_s = 98\pm5$ kA/m (98±5 emu/cm^3). The small value of g can be explained by the mean field theory for ferrimagnets (cp. section 2.4). Close to the magnetic compensation point the g-factor should (in the ideal case) tend to zero. This effect has been experimentally verified for a CoGd sample in chapter 7. In Fig. 7.6 c) the effective g-factor g_{eff} as a function of the temperature as extracted from FMR data of a CoGd sample is plotted. On the left side of the peak the g-factor slightly decreases to values of about 1.3. The FeGd sample investigated in this chapter is close to the magnetic compensation point which is supported by the SQUID data.

From the SQUID measurements and the estimated heating of 20 K due to the laser M_s is expected to be around 90 kA/m (90 emu/cm^3). The fit value of $M_s = 98 \pm 5$ kA/m (98 ± 5 emu/cm^3) is slightly higher but nevertheless reasonable within the accuracy of the fit. This can be an indicator that the heating due to the laser is a bit lower than the estimated 20 K. Furthermore the inhomogeneity of the sample can be a reason for the slightly larger value. However, the fit quality using the Kittel formula for pure in-plane magnetization (2.19) emphasizes the dominance of the in-plane component as indicated by the SQUID measurements.

Temperature dependent in-plane FMR observations reveal that the magnetization dynamics is dominated by the applied magnetic field as the resonance field changes slightly from 414 to 430 kA/m (5200 - 5400 Oe). This is expected since the magnetization is very small. At the same time the linewidth decreases from 30.2 to 25.4 kA/m (380 to 320 Oe). The corresponding damping parameter α is evaluated using equation (4.3) is depicted in Fig. 8.4 a) and has a value of $\alpha = 0.033 \pm 0.003$. Temperature dependent TR-MOKE measurements reveal no significant shift of the precessional frequency f as function of the temperature. $f = 7.6 \pm 0.03$ GHz is found. α is about 0.120 ± 0.004, which is the same value as for the field dependent measurements. α was evaluated according to equation (4.18) and is shown as a function of the applied in-plane magnetic field in Fig. 8.4 b). α is presented for applied fields of ≥ 39.8 kA/m (500 Oe). Only for those fields multiple oscillations of the magnetization within the length of the delay are obtained (cp. Fig. 8.2) and the determination of the decay time τ of the uniform mode is reliable. For smaller fields the uncertainties involved in the fits become larger. α is found to be constant. For both, the evaluation of the FMR and TR-MOKE measurement, the g-factor g=1.5 according to the Kittel fit is taken into account.

The values for the damping parameter α are about three times larger in the TR-MOKE experiments compared to FMR. Numerical solutions of the LLG-equation support the findings (cp. Fig. 8.4 c)). Various mechanism can contribute to the observed increased damping in the TR-MOKE experiments. Most likely a sample position with higher local damping due to sample inhomogeneities was probed. Furthermore, the observed movement of the pump beam due to the

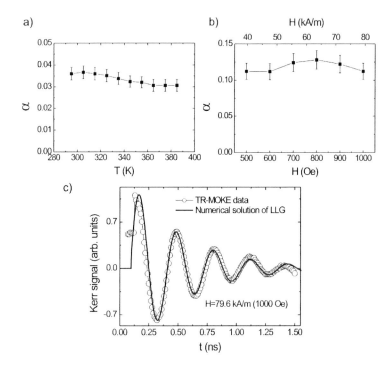

Figure 8.4: a) Damping parameter α as measured by in-plane FMR as a function of the temperature T, $\alpha = 0.033 \pm 0.003$. b) α as evaluated from TR-MOKE experiments as a function of applied in-plane magnetic field. Again the value is constant, $\alpha = 0.120 \pm 0.004$. c) Numerical solution of the LLG-equation with $\alpha = 0.12$.

mechanical time delay can significantly change the magnitude of the excitation and lead to a reduction of the decay time τ, which pretends an increased damping (cp. eqn. (4.18)). Radial damping via spin waves with wave vectors $k \neq 0$, that are excited through inhomogeneities during the optical excitation, can also increase α in the TR-MOKE experiments. Note that a big advantage of FMR measurements is that the damping parameter α is 'directly' measured as no material specific parameters except the g-factor are required to determine

α (cp. eqn. (4.3)). In contrast another material specific parameter, namely the saturation magnetization M_s, is also necessary and very important for the extraction of α from the TR-MOKE data (cp. eqn. (4.18). Hence a uncertainty in M_s can lead to a (large) error of the damping parameter α. Uncertainties of M_s can arise from the sample inhomogeneities and the unknown laser heating in TR-MOKE experiments. Thus an influence on the measured damping parameter α for FeGd can also not be excluded in the TR-MOKE experiment presented here.

Besides the Kerr signal the temporal behavior of the reflectivity is of special interest as the reflectivity is a measure for the sample temperature [122]. In Fig. 8.5 a) two signal traces for fluences of 15.6 mJ/cm^2 and 43.0 mJ/cm^2 are depicted, respectively. As expected, the amplitude of the reflectivity scales nearly linearly with the fluence, i. e. the excitation. The decay time τ_r of the reflectivity is fitted as introduced in chapter 5, equation (5.3), using the relation

$$y(t) = A \cdot \exp(t/\tau_r) + y_0. \tag{8.2}$$

The evaluation yields a decay time τ_r of about 0.35 ± 0.2 ns. No dependence of τ_r on the fluence can be observed (cp. Fig. 8.5 b)). A detailed examination shows that the background decay time τ_b and the reflectivity decay time τ_r are both approximately 0.35 ± 0.2 ns. Therefore the magnetic background is directly correlated to the temperature of the sample which is expected as the exponential magnetic background is due to the temperature dependence of transient effective magnetic field $\boldsymbol{H}_{eff}*$ (cp. subsection 2.5.3 and Fig. 2.15). This is also in agreement with findings on NiFe which are presented in chapter 5.

8.3 Conclusion

The multilayered FeGd sample is well suited for the investigation by TR-MOKE as the canted magnetic state allows an easy triggering of magnetization dynamics. The experiments show that the magnetization is very small and mainly oriented in-plane in a temperature range from RT up to 350 K. We do not observe a temperature induced SRT as in the X-PEEM experiments. This is attributed to the fact that the sample is very inhomogeneous. Consequently only data obtained at one sample position can be correlated. In agreement with the

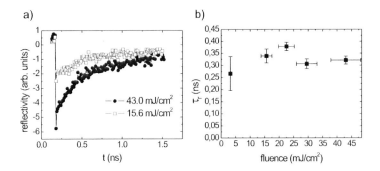

Figure 8.5: a) Reflectivity as a a function of the pump fluence. A clear dependence of the signal amplitude to the excitation is evident. In both cases the signal describes an exponential decay. b) The decay time τ_r is constant as a function of the pump fluence.

mean-field model for ferrimagnets we find g=1.5±0.05. The damping parameter α is found to be $\alpha = 0.033\pm0.003$ in FMR, and $\alpha = 0.120\pm0.004$ in TR-MOKE. The damping parameter α does not significantly change as a function of field or temperature. The discrepancy between FMR and TR-MOKE is due to the very localized nature of the TR-MOKE measurements. The sample inhomogeneity, spin wave excitations and a changing amplitude of pumping in the TR-MOKE experiment can lead to the observed increased α.

The results could not verify the X-PEEM results. In all experiments we found decay times of the uniform precession, τ, and the reflectivity, τ_r, below 1 ns. Compared to results on similar systems like CoGd this is reasonable. As the sample is very inhomogeneous it is hard to compare the experiments directly. The SRT which was found in the X-PEEM measurements, may be the reason for the different magnetization dynamics. Well defined and homogenous samples are necessary for further investigations.

9 Summary and Outlook

9.1 Summary

In this thesis the magnetization dynamics in rare-earth doped ferromagnetic materials was investigated using all-optical TR-MOKE, VNA-FMR and conventional FMR experiments. First, the well known ferromagnet $Ni_{80}Fe_{20}$ was used since it is a well studied ferromagnet and has a small intrinsic magnetic damping. Subsequently, experiments with RE-doped $Ni_{80}Fe_{20}$ samples, which are ferrimagnets, were performed. The measurements have shown that TR-MOKE and (VNA-)FMR experiments are well suited techniques to investigate the magnetization dynamics of ferrimagnets. After that, experiments on two particular ferrimagnets, CoGd and FeGd, were carried out. The additional damping for both alloys due to the spin-orbit coupling of Gd is minimized as Gd has essentially zero orbital momentum and thus the additional contribution to damping via the spin-orbit coupling is negligible. For CoGd it was possible to investigate the magnetization dynamics in the vicinity of the compensation points.

In chapter 5 measurements on the ferromagnet $Ni_{80}Fe_{20}$ were discussed. The extracted damping parameter α of the uniform precession mode from the TR-MOKE experiments yielded a 2-3 times larger values for the damping parameter α compared to the one obtained by FMR, which is in agreement with literature values. The pretended increased damping in the TR-MOKE measurements was mainly attributed to a displacement of the pump beam with respect to the probe beam during the measurements due to the mechanical delay stage. The damping parameter α was found to be constant as a function of the applied field and the laser fluence. Beside the uniform mode we detected the signature of a first order PSSW in VNA-FMR and FMR. The increased damping for low frequencies in VNA-FMR was explained by a model of Counil et. al [129]. The extrinsic damping contribution was found to be very small in $Ni_{80}Fe_{20}$. The intrinsic α

is evaluated by FMR at a frequency of 21.9 GHz to be $\alpha = 0.0076 \pm 0.0004$. Due to measurements of the precessional frequency as a function of the applied magnetic field the g-factor and the exchange stiffness constant were obtained.

In chapter 6 the results of Gd, Ho and Dy doped $Ni_{80}Fe_{20}$ samples were presented. The measurements showed that RE dopants strongly influence the magnetization dynamics of NiFe via the SO-coupling. In agreement with considerations based on Hund's rules the influence of Gd on the damping was found to be minimal. Ho and Dy led to large changes of the damping parameter α even for small RE contents. In all measurements we find a linear dependence of the damping parameter α on the RE content. The additional damping contribution is larger for Dy than for Ho. This was not expected. Considering all results we do not have an explanation for this finding. Extrinsic damping was excluded as reason for the samples with low doping. A structural phase transition, sample inhomogeneities and defects may provide an explanation for samples with higher doping level. So far a detailed structural analysis is missing.

In chapter 7 the magnetization dynamics of the ferrimagnet CoGd were discussed. The magnetic damping and resonance frequency in the vicinity of the magnetization and angular momentum compensation point were studied. In doing so a large increase in precessional frequency and damping for the total magnetization was observed. This is in agreement with the mean-field model, that was introduced in section 2.4. CoGd allows one to tune the precessional frequency f and the damping parameter α_{eff} by means of the sample composition and the temperature of the sample. The TR-MOKE study showed that an all-optical control of the magnetization dynamics in the vicinity of the compensation points may be feasible.

In chapter 8 another ferrimagnet, FeGd, was investigated in detail. The FeGd was a multilayered sample, not an alloy. The sample was well suited for the investigation by TR-MOKE since its canted magnetic state allows one to easily trigger magnetization dynamics. No temperature induced SRT was found for this sample in contrast to the X-PEEM experiments. In agreement with the mean-field model for ferrimagnets a small g-factor was found. The damping parameter α did not significantly change as a function of the applied magnetic field or temperature. In all experiments we found decay times of the uniform precession and the reflectivity below 1 ns. The results for the decay times obtained by TR-MOKE and FMR differ from the X-PEEM results at least by a factor of 2. The sample turned out to be very inhomogeneous.

9.2 Outlook

The structure of the RE-doped NiFe samples seems to be very important for the understanding of the damping behavior. Hence, a detailed structural analysis of the investigated samples is required. This may provide an explanation for the larger damping contribution observed for Dy doping compared to Ho doping. The analysis can also verify if the structural phase transition from the polycrystalline to the amorphous state is present in the samples for higher dopant concentrations.

A further step can be taken to investigate different RE doped ferromagnets since so far mainly NiFe was used. These experiments can verify whether the contributed damping due to the RE doping is the same in various ferromagnetic materials.

An important experimental point is that the motion of the pump beam in the TR-MOKE experiments as a function of time needs to be suppressed in order to obtain reliable results for the damping parameter α. First experiments with an active beam stabilizer system have been recently performed and are very promising.

Appendix A - Sample Properties

NiFe ($Ni_{80}Fe_{20}$) and RE-doped NiFe

The saturation magnetization M_s and the effective g-factor g_{eff} are important parameters for the evaluation of the magnetization dynamics of NiFe and RE-doped NiFe samples in chapters 5 and 6. The values that were measured via Vibrating Sample Magnetometery (VSM) and angle dependent FMR, respectively, are summarized in Table A.1. For some samples with higher doping, no g_{eff} factors could be evaluated. Thus g_{eff} was chosen to be 2.12, the value for pure NiFe, for the calculations of samples, for which no experimental determination of g_{eff} was possible. All samples consist of 30 nm thick NiFe sputtered onto glass and are capped with 5 nm Ta to prevent them from oxidation.

The behavior of M_s and g_{eff} as a function of the RE content can be understood based on the theory of ferrimagnetism and the simple mean field model presented in section 2.4 as NiFeRE samples are ferrimagnets. The reduction of M_s with increasing RE content is due to the antiferromagnetic coupling of the NiFe sublattice to the RE sublattice. Note that this is only valid as long as the dopant concentrations are low and the samples are below the magnetization compensation point. As can be seen in Figure A.1, M_s is a linear function of the rare earth concentration.

Temperature dependent VSM measurements on the NiFeDy samples with Dy-contents varying between 5% and 20% show that a composition of about 20% Dy yields the magnetic compensation point at RT. Similar compositions are expected for Gd and Ho due to the similar M_s versus RE-content relation. Therefore, the increase of g_{eff} for NiFeGd (cp. Tab. A.1) can very likely be understood by the mean field model. With increasing Gd content the sample composition gets closer to the point at which the magnetic compensation temperature T_M is near RT. Towards the compensation point an increasing g_{eff} is expected.

sample	M_s in kA/m (emu/cm^3)	g_{eff}
NiFe (Ni$_{80}$Fe$_{20}$)	807	2.12
NiFeGd5	621	2.12
NiFeGd10	419	2.17
NiFeGd15	212	2.27
NiFeHo2	751	2.12
NiFeHo4	621	2.12
NiFeHo6	567	-
NiFeHo8	441	-
NiFeDy2.5	685	2.13
NiFeDy5	559	-

Table A.1: Measured properties of the RE-doped NiFe samples. The number in the sample name corresponds to the RE-content in % (accuracy $\pm1\%$, determined by RBS). M_s was determined using VSM ($\Delta M_S/M_s = \pm4\%$). The effective g-factor was evaluated by M. Kießling [130] via angle dependent FMR ($\Delta g \leq 0.05$).

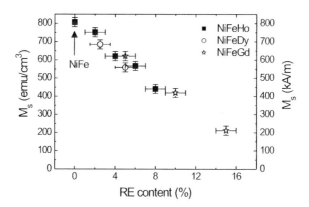

Figure A.1: M_s versus RE content. A linear dependence of M_s on the RE content can be seen. The decay is very similar for all dopants.

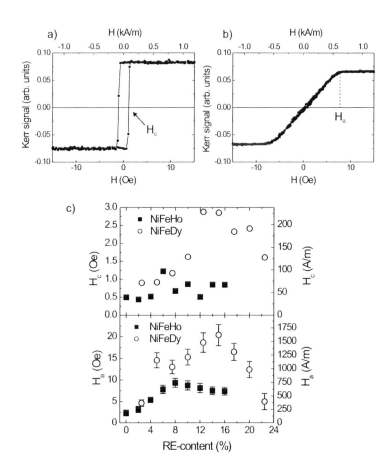

Figure A.2: a) shows an easy b) a hard axis hysteresis loop for NiFeHo8 as measured by MOKE. From the easy axis loop the coercive field H_c, from the hard axis loop the anisotropy field H_a is extracted. c) Results for H_c and H_a for the NiFeHo and NiFeDy samples as extracted from hysteresis loops. For H_c the error bars ($\Delta H_c < 0.5$ Oe) are omitted for clarity.

Fig. A.2 a) shows a typical easy axis hysteresis loop for NiFeHo8 from which H_c is determined and b) a hard axis hysteresis loop from which the anisotropy field H_a, i. e. the saturation field for the hard axis of the sample is determined. Fig. A.2 c) summarizes the results for H_c and H_a for the NiFeHo and NiFeDy samples. The results reveal that the anisotropy contributions for the data evaluation in chapter 5 and 6 can be neglected as the anisotropy fields are small compared to the typical applied magnetic fields of some hundred kA/m (Oe). See [130] for further details.

Another interesting aspect can be seen in Fig. A.2 c). The anisotropy field H_a for both dopants increases with increasing RE content until a maximum is reached for 8% Ho and 15% Dy, respectively. If the RE concentration is further increased the anisotropy field decreases. Bailey et al. [146] observed similar results in RE doped Py films and attributed this to a structural phase transition from the polycrystalline to the amorphous state.

Appendix B - Publications

Magnetization dynamics of the ferrimagnet CoGd near the compensation of magnetization and angular momentum,
M. Binder, A. Weber, O. Mosendz, G. Woltersdorf, M. Izquierdo, I. Neudecker,
J. R. Dahn, T. D. Hatchard, J.-U. Thiele, C. H. Back and M. R. Scheinfein,
Phys. Rev. B **74**, (134404), 2006

Slow recovery of magnetic anisotropy following ultrafast optical excitation of a spin-reorientation transition,
T. Eimüller, A. Scholl, B. Ludescher, M. Binder, G. Schütz, C. H. Back and
J. Stöhr,
in preparation, 2006

Effect of Rare Earth Dopants in Permalloy Films and Multilayers,
G. Woltersdorf, J.-U. Thiele, M. Schabes, G. Meyer, M. Kiessling, M. Binder
and C. H. Back,
Verhandl. DPG (VI) 41, 1/2006 (2006), ISSN 0420-0195

Magnetization Dynamics of the Ferrimagnet CoGd near the compensation point,
M. Binder, A. Weber, O. Mosendz, G. Woltersdorf, M. Izquierdo, I. Neudecker,
J. R. Dahn, T. D. Hatchard, J.-U. Thiele, C. H. Back and M. R. Scheinfein,
Verhandl. DPG (VI) 41, 1/2006 (2006), ISSN 0420-0195

Untersuchungen zur Lagengüte an TMR-Schichtsystemen,
M. Binder and J. Zweck,
Verhandl. DPG (VI) 37, 1/2002 (2002), ISSN 0420-0195

Bibliography

[1] A. S. Arrott. *Magnetism in SI-Units and Gaussian Units* In: *Ultrathin Magnetic Structures I,* edited by J.A.C. Bland and B. Heinrich. Springer Verlag (1994).

[2] M. N. Baibich, J. M. Broto, A. Fert, F. N. van Dau, F. Petroff, P. Eitenne, G. Creuzet, A. Friederich, J. Chazelas. *Giant magnetoresistance of (001)Fe/(001)Cr magnetic superlattices.* Phys. Rev. Lett., **61**, 2472 (1988).

[3] G. Binasch, P. Grünberg, F. Saurenbach, W. Zinn. *Enhanced magnetoresistance in layered magnetic structures with antiferromagnetic interlayer exchange.* Phys. Rev. B, **39**, 4828 (1989).

[4] J. S. Moodera, L. R. Kinder, T. M. Wong, R. Meservey. *Large magnetoresistance at room temperature in ferromagnetic thin film tunnel junctions.* Phys. Rev. Lett., **74**, 3273 (1995).

[5] S. A. Wolf, D. D. Awschalom, R. A. Buhrmann, J. M. Daughton, S. von Molnár, M. L. Roukes, A. Y. Chtchelkanova, D. M. Treger. *Spintronics: A spinbased electronic vision for the future.* Science, **294**, 1488 (2001).

[6] G. A. Prinz. *Magnetoelectronics.* Science, **282**, 1660 (1998).

[7] S. Tehrani, E. Chen, M. Durlam, M. Deherrera, J. M. Slaughter, J. Shi, G. Kerszykowski. *High density submicron magnetoresistive random access memory (invited).* J. Appl. Phys., **85**, 5822 (1999).

[8] Y. Tserkovnyak, A. Brataas, G. E. W. Bauer. *Enhanced Gilbert damping in thin ferromagnetic films.* Phys. Rev. Lett., **88**, 117601 (2002).

[9] B. B. Maranville, J. Mallett, T. P. Moffat, R. D. McMichael, A. P. Chen, J. W. F. Egelhoff. *Effect of conformal roughness on ferromagnetic resonance linewidth in thin permalloy films.* J. Appl. Phys., **97**, 10A721 (2005).

[10] V. Kamberský. *On the Landau-Lifshitz relaxation in ferromagnetic metals.* Can. J. Phys., **48**, 2906 (1970).

[11] S. G. Reidy, L. Cheng, W. E. Bailey. *Dopants for independent control of precessional frequency and damping in $Ni_{81}Fe_{19}$ (50 nm) thin films.* Appl. Phys. Lett., **82**, 1254 (2002).

[12] J. P. Nibarger, R. Lopusnik, T. J. Silva. *Damping as a function of pulsed field amplitude and bias field in thin film permalloy.* Appl. Phys. Lett., **82**, 2112 (2003).

[13] M. Binder, A. Weber, O. Mosendz, G. Woltersdorf, M. Izquierdo, I. Neudecker, J. R. Dahn, T. D. Hatchard, J.-U. Thiele, C. H. Back, M. R. Scheinfein. *Magnetization dynamics of the ferrimagnet CoGd near the compensation of magnetization and angular momentum.* Phys. Rev. B, **74**, 134404 (2006).

[14] C. Kittel. *Physical theory of ferromagnetic domains.* Rev. Mod. Phys., **21**, 541 (1949).

[15] S. Blundell. *Magnetism in Condensed Matter.* Oxford University Press, Oxford (2001).

[16] H. Haken, H. C. Wolf. *Atom- und Quantenphysik.* Springer Verlag, Berlin (1996).

[17] W. F. Brown. *Micromagnetics.* Interscience Publishers, New York (1963).

[18] J. Miltat, G. Albuquerque, A. Thiaville. *An Introduction to Micromagnetics in the Dynamic Regime* In: *Spin Dynamics in Confined Magnetic Structures I,* edited by B. Hillebrands and K. Ounadjela. Springer Verlag (2002).

[19] A. Hubert, R. Schäfer. *Magnetic Domains - The Analysis of Magnetic Microstructures.* Springer Verlag, Berlin (2000).

[20] A. Aharoni. *Introduction to the theory of ferromagnetism*. Oxford University Press, New York (2000).

[21] C. Kittel. *Einführung in die Festkörperphysik*. Oldenbourg Verlag, München (1999).

[22] L. D. Landau, E. Lifshitz. *On the theory of dispersion of magnetic permeability in ferromagnetic bodies*. Physik Z. Sowjetunion, **8**, 153 (1935).

[23] J. A. Osborn. *Demagnetizing factors of the general ellipsoid*. Phys. Rev., **67**, 351 (1945).

[24] T. L. Gilbert. *A Lagrangian formulation of the gyromagnetic equation of the magnetization field*. Phys. Rev., **100**, 1243 (1955).

[25] T. L. Gilbert. *Formulation, foundations and applications of the phenomenological theory of ferromagnetism*. PhD thesis, Illinois Institiute of Technology (1956).

[26] J. C. Mallinson. *On damped gyro-magnetic precession*. IEEE Trans. Magn., **23**, 2003 (1987).

[27] S. E. Russek, R. D. McMichael, M. J. Donahue, S. Kaka. *High Speed Switching and Rotational Dynamics in Small Magnetic Thin Film Devices* In: *Spin Dynamics in Confined Magnetic Structures II,* edited by B. Hillebrands and K. Ounadjela. Springer Verlag (2003).

[28] M. van Kampen. *Ultrafast spin dynamics in ferromagnetic materials*. PhD thesis, University of Eindhoven, The Netherlands (2003).

[29] F. Bloch. *Zur Theorie des Ferromagnetismus*. Z. Phys., **61**, 206 (1930).

[30] D. D. Stancil. *Theory of Magnetostatic Waves*. Springer Verlag, New York (1993).

[31] S. O. Demokritov, B. Hillebrands, A. N. Slavin. *Brillouin light scattering studies of confined spin waves: Linear and nonlinear confinement*. Phys. Rep., **348**, 441 (2001).

[32] R. Arias, D. L. Mills. *Extrinsic contributions to the ferromagnetic resonance response of ultrathin films*. Phys. Rev. B, **60**, 7395 (1999).

[33] C. Herring, C. Kittel. *On the theory of spin waves in ferromagnetic media.* Phys. Rev., **81**, 869 (1951).

[34] B. A. Kalinikos, A. N. Slavin. *Theory of dipole-exchange spin wave spectrum for ferromagnetic films with mixed exchange boundary conditions.* J. Phys. C, **19**, 7013 (1986).

[35] S. O. Demokritov, B. Hillebrands. *Spinwaves in Laterally Confined Magnetic Structures* In: *Spin Dynamics in Confined Magnetic Structures I,* edited by B. Hillebrands and K. Ounadjela. Springer Verlag (2002).

[36] S. O. Demokritov. *Dynamic eigen-modes in magnetic stripes and dots.* J. Phys.: Condens. Matter, **15**, 2575 (2003).

[37] A. H. Morrish. *The Physical Principles of Magnetism.* John Wiley & Sons, New York (1966).

[38] B. Heinrich, R. Urban, G. Woltersdorf. *Magnetic relaxation in metallic films: single and multilayer structures.* J. Appl. Phys., **91**, 7523 (2002).

[39] B. Heinrich. *Spin Relaxation in Magnetic Metallic Layers and Multilayers* In: *Ultrathin Magnetic Structures III,* edited by J. A. C. Bland and B. Heinrich. Springer Verlag (2005).

[40] G. Woltersdorf. *Spin-pumping and two-magnon scattering in magnetic multilayers.* PhD thesis, Simon Fraser University, Burnaby, Canada (2004).

[41] E. A. Turov. *Features of Ferromagnetic Resonance in Metals* In: *Ferromagnetic Resonance,* edited by J. A. C. Bland and B. Heinrich. Pergamon Press (1966).

[42] V. Kamberský. *On ferromagnetic resonance damping in metals.* Czech. J. Phys. B, **26**, 1366 (1976).

[43] B. Heinrich, D. Fraitová, V. Kamberský. *The influence of s-d-exchange on relaxation of magnons in metals.* Phys. Stat. Sol., **23**, 501 (1967).

[44] J. Kunes, V. Kamberský. *First-principles investigation of the damping of fast magnetization precession in ferromagnetic 3d metals.* Phys. Rev. B, **65**, 212411 (2002).

[45] V. Kamberský. *FMR linewidth and disorder in metals.* Czech. J. Phys. B, **34**, 1111 (1984).

[46] V. Kamberský, J. F. Cochran, J. M. Rudd. *Anisotropic low-temperature FMR linewidth in nickel and the theory of 'anomalous' damping.* J. Magn. Magn. Mat., **104**, 2089 (1992).

[47] H. Suhl. *Theory of the magnetic damping constant.* IEEE Trans. Magn., **34**, 1834 (1998).

[48] E. Schlömann. *Spin-wave analysis of ferromagnetic resonance in polycrystalline ferrites.* J. Phys. Chem. Solids, **6**, 242 (1958).

[49] R. C. LeCraw, E. G. Spencer, C. S. Porter. *Ferromagnetic resonance line width in yttrium iron garnet single crystals.* Phys. Rev., **110**, 1311 (1958).

[50] M. J. Hurben, D. R. Franklin, C. E. Patton. *Angle dependence of the ferromagnetic resonance linewidth in easy-axis and easy-plane single crystal hexagonal ferrite disks.* J. Appl. Phys., **81**, 7458 (1997).

[51] B. Heinrich, J. F. Cochran, R. Hasegawa. *FMR linebroadening in metals due to two-magnon scattering.* J. Appl. Phys., **57**, 3690 (1985).

[52] C. E. Patton, C. H. Wilts, F. B. Humphrey. *Relaxation processes for ferromagnetic resonance in thin films.* J. Appl. Phys., **38**, 1358 (1967).

[53] C. Józsa. *Optical detection of the magnetization precession - choreography on a sub-nanosecond timescale.* PhD thesis, University of Eindhoven, The Netherlands (2005).

[54] O. Madelung (ed.). *Landolt-Börnstein - Magnetic Properties of Metals, New Series III/19 g.* Springer Verlag, Berlin (1988).

[55] A. Gangulee, R. Taylor. *Mean field analysis of the magnetic properties of vapor deposited amourphous Fe-Gd thin films.* J. Appl. Phys., **49**, 1762 (1978).

[56] M. Mansuripur. *The Physical Principles of Magneto-optical Recording.* Cambridge University Press, Cambridge (1995).

[57] R. K. Wangsness. *Sublattice effects in magnetic resonance.* Phys. Rev., **91**, 1085 (1953).

[58] J. Kaplan, C. Kittel. *Dynamic eigen-modes in magnetic stripes and dots.* J. Chem. Phys., **21**, 760 (1953).

[59] E. Beaurepaire, J.-C. Merle, A. Daunois, J.-Y. Bigot. *Ultrafast spin dynamics in ferromagnetic nickel.* Phys. Rev. Lett., **76**, 4250 (1996).

[60] W. S. Fann, R. Storz, H. W. K. Tom, J. Bokor. *Direct measurement of nonequilibrium electron-energy distributions in subpicosecond laser-heated gold films.* Phys. Rev. Lett., **68**, 2834 (1992).

[61] S. I. Anisimov, B. L. Kapeliovich, T. L. Perelman. *Electron emission from metal surfaces exposed to ultrashort laser pulses.* Sov. Phys. JETP, **39**, 375 (1974).

[62] H.-S. Rhie, H. A. Dürr, W. Eberhardt. *Femtosecond Electron and Spin Dynamics in Ni/W(110) Films.* Phys. Rev. Lett., **90**, 247201 (2003).

[63] R. Knorren, K. H. Bennemann, R. Burgermeister, M. Aeschlimann. *Dynamics of excited electrons in copper and ferromagnetic transition metals: theory and experiment.* Phys. Rev. B, **61**, 9427 (2000).

[64] B. Koopmans, M. van Kampen, J. T. Kohlhepp, W. J. M. de Jonge. *Ultrafast magneto-optics in nickel: magnetism or optics?* Phys. Rev. Lett., **85**, 844 (2000).

[65] R. H. M. Groeneveld, R. Sprik, A. Lagendijk. *Femtosecond spectroscopy of electron-electron and electron-phonon energy relaxation in Ag and Au.* Phys. Rev. B, **51**, 11433 (1995).

[66] P. J. van Hall. *Ultrafast processes in Ag and Au: A Monte Carlo study.* Phys. Rev. B, **63**, 104301 (2001).

[67] W. S. Fann, R. Storz, H. W. K. Tom, J. Bokor. *Electron thermalization in gold.* Phys. Rev. B, **46**, 13592 (1992).

[68] C. K. Sun, F. Vallee, L. H. Acioli, E. P. Ippen, J. G. Fulimoto. *Femtosecond-tunable measurement of electron thermalization in gold.* Phys. Rev. B, **50**, 15337 (1994).

[69] D. Bejan, G. Raseev. *Nonequilibrium electron distribution in metals.* Phys. Rev. B, **55**, 4250 (1997).

[70] N. del Fatti, C. Voisin, M. Achermann, S. Tzortzakis, D. Christofilos, F. Vallee. *Nonequilibrium electron dynamics in noble metals.* Phys. Rev. B, **61**, 16956 (2000).

[71] G. Tas, H. J. Maris. *Electron diffusion in metals studied by picosecond ultrasonics.* Phys. Rev. B, **49**, 15046 (1994).

[72] J. Hohlfeld, S.-S. Wellershoff, J. Güdde, U. Conrad, V. Jahnke, E. Matthias. *Electron and lattice dynamics following optical excitation of metals.* Chem. Phys., **251**, 237 (2000).

[73] B. Koopmans. *Laser Induced Magnetization Dynamics* In: *Spin Dynamics in Confined Magnetic Structures II,* edited by B. Hillebrands and K. Ounadjela. Springer Verlag (2003).

[74] B. Koopmans, J. J. M. Ruigrok, F. D. Longa, W. J. M. de Jonge. *Unifying ultrafast magnetization dynamics.* Phys. Rev. Lett., **95**, 267207 (2005).

[75] M. Cinchetti, M. S. Albaneda, D. Hoffmann, T. Roth, J.-P. Wüstenberg, M. Krauß, O. Andreyev, H. C. Schneider, M. Bauer, M. Aeschlimann. *Spin-flip processes and ultrafast magnetization dynamics in Co: Unifying the microscopic and macroscopic view of femtosecond magnetism.* Phys. Rev. Lett., **97**, 177201 (2006).

[76] A. Vaterlaus, T. Beutler, D. Guarisco, M. Lutz, F. Meier. *Spin-lattice relaxation in ferromagnets studied by time-resolved spin-polarized photoemission.* Phys. Rev. B, **46**, 5280 (1992).

[77] A. Vaterlaus, T. Beutler, F. Meier. *Spin lattice relaxation time of ferromagnetic gadolinium determined with time-resolved spin-polarized photoemission.* Phys. Rev. Lett., **67**, 3314 (1991).

[78] W. Hübner, K. H. Bennemann. *Simple theory for spin-lattice relaxation in metallic rare-earth ferromagnets.* Phys. Rev. B, **53**, 3422 (1996).

[79] J. Hohlfeld, E. Matthias, R. Knorren, K. H. Bennemann. *Nonequilibrium magnetization dynamics of nickel.* Phys. Rev. Lett., **78**, 4861 (1997).

[80] J. Güdde, U. Conrad, V. Jähnke, J. Hohlfeld, E. Matthias. *Magnetization dynamics of Ni and Co films on Cu(001) and of bulk nickel surfaces.* Phys. Rev. B, **59**, R6608 (1999).

[81] J. Hohlfeld, J. Güdde, U. Conrad, O. Dühr, G. Korn, E. Matthias. *Ultrafast magnetization dynamics of nickel.* Appl. Phys. B, **68**, 505 (1999).

[82] U. Conrad, J. Güdde, V. Jähnke, E. Matthias. *Ultrafast electron and magnetization dynamics of thin Ni and Co films on Cu(001) observed by time-resolved SHG.* Appl. Phys. B, **68**, 511 (1999).

[83] A. Scholl, L. Baumgarten, R. Jacquemin, W. Eberhardt. *Ultrafast spin dynamics of ferromagnetic thin films observed by fs spin-resolved two-photon photoemission.* Phys. Rev. Lett, **79**, 5146 (1997).

[84] J. Hohlfeld, T. Gerrits, M. Bilderbeek, T. Rasing, H. Awano, N. Ohta. *Fast magnetization reversal of GdFeCo induced by femtosecond laser pulses.* Phys. Rev. B, **65**, 012413 (2001).

[85] W. Hübner, G. P. Zhang. *Ultrafast spin dynamics in nickel.* Phys. Rev. B, **58**, R5920 (1998).

[86] G. P. Zhang, W. Hübner. *Laser-induced ultrafast demagnetization in ferromagnetic metals.* Phys. Rev. Lett., **85**, 3025 (2000).

[87] G. P. Zhang, W. Hübner. *Femtosecond spin dynamics in the time domain.* J. Appl. Phys., **85**, 5657 (1999).

[88] H. Vogel. *Gerthsen Physik.* Springer Verlag, Berlin (1995).

[89] A. Weber. *Ultrafast Magnetization Dynamics induced by Femtosecond Laser Pulses.* Master's thesis, University of Regensburg, Regensburg, Germany (2004).

[90] P. Hansen, C. Clausen, G. Much, M. Rosenkranz, K. Witter. *Magnetic and magneto-optical properties of rare-earth transition metal alloys containing Gd, Tb, Fe, Co.* J. Appl. Phys., **66**, 756 (1989).

[91] L. C. Feldman, J. W. Mayer. *Fundamentals of surface and thin film analysis.* North-Holland, New York (1986).

[92] J. R. Dahn, S. Trussler, T. D. Hatchard, A. Bonakdarpour, J. R. Mueller-Neuhaus, K. C. Hewitt, M. Fleischauer. *Economical sputtering system to produce large-size composition-spread libraries having linear and orthogonal stochiometry variations.* Chem. Mater., **14**, 3519 (2002).

[93] D. A. R. Barkhouse, A. Bonakdarpour, M. Fleischauer, T. D. Hatchard, J. R. Dahn. *A combinatorial sputtering method to prepare a wide range of A/B artificial superlattice structures on a single substrate.* J. Magn. Magn. Mat., **261**, 399 (2003).

[94] A. Bonakdarpour, K. C. Hewitt, T. D. Hatchard, M. D. Fleischauer, J. R. Dahn. *Combinatorial synthesis and rapid characterization of $Mo_{1-x}Sn_x$ $(0 \leq x \leq 1)$ thin films.* Thin Solid Films, **440**, 11 (2003).

[95] B. Heinrich. *Radio Frequency Techniques - Ferromagnetic Resonance in Ultrathin Film Structures* In: *Ultrathin Magnetic Structures II,* edited by B. Heinrich and J. A. C. Bland. Springer Verlag, Berlin (1994).

[96] Z. Celinski, K. B. Urquhart, B. Heinrich. *Using ferromagnetic resonance to measure the magnetic moments of ultrathin films.* J. Magn. Magn. Mat., **166**, 6 (1997).

[97] B. Heinrich, J. F. Cochran. *Ultrathin metallic magnetic films: magnetic anisotropies and exchange interaction.* Adv. Phys., **42**, 523 (1993).

[98] C. Kittel. *Excitation of spin waves in a ferromagnet by a uniform rf field.* Phys. Rev., **110**, 1295 (1958).

[99] Z. Celinski, B. Heinrich. *Ferromagnetic resonance linewidth of Fe ultrathin films grown on a bcc Cu substrate.* J. Appl. Phys., **70**, 5935 (1991).

[100] M. Bailleul, D. Olligs, C. Fermon. *Propagating spin wave spectroscopy in a permalloy film: A quantitative analysis.* Appl. Phys. Lett., **83**, 972 (2003).

[101] M. Bailleul, D. Olligs, C. Fermon. *Micromagnetic phase transitions and spin wave excitations in a ferromagnetic stripe.* Phys. Rev. Lett., **91**, 137204 (2003).

[102] I. Neudecker. *Magnetization Dynamics of Confined Ferromagnetic Systems.* PhD thesis, Universität Regensburg, Regensburg, Germany (2006).

[103] I. Neudecker, G. Woltersdorf, B. Heinrich, T. Okuno, G. Gubbiotti, C. H. Back. *Comparison of frequency, field, and time domain ferromagnetic resonance methods.* J. Magn. Magn. Mat., **307**, 148 (2006).

[104] B. Hillebrands. *Brillouin light scattering from layered magnetic structures* In: *Light Scattering in Solids VII*, edited by M. Cardona and G. Güntherodt. Springer Verlag, Berlin (2000).

[105] J. R. Sandercock. *Trends in Brillouin light scattering: Studies of Opaque Materials, Supported Films, and Central Modes* In: *Light Scattering in Solids III*, edited by M. Cardona and G. Güntherodt. Springer Verlag, Berlin (1979).

[106] G. Gubbiotti, G. Carlotti, T. Okuno, T. Shinjo, F. Nizzoli, R. Zivieri. *Brillouin light scattering investigation of dynamic spin modes confined in cylindrical permalloy dots.* Phys. Rev. B, **68**, 184409 (2003).

[107] L. Giovannini, F. Montoncello, F. Nizzoli, G. Gubbiotti, G. Carlotti, T. Okuno, T. Shinjo, M. Grimsditch. *Spin excitations of nanometric cylindrical dots in vortex and saturated magnetic states.* Phys. Rev. B, **70**, 172404 (2004).

[108] K. Perzlmaier, M. Buess, C. H. Back, V. E. Demidov, B. Hillebrands, S. O. Demokritov. *Spin-wave eigenmodes of permalloy squares with a closure domain structure.* Phys. Rev. Lett., **94**, 057202 (2005).

[109] A. N. Slavin, S. O. Demokritov, B. Hillebrands. *Nonlinear Spinwaves in One- and Two-Dimensional Magnetic Waveguides* In: *Spin Dynamics in Confined Magnetic Structures I*, edited by B. Hillebrands and K. Ounadjela. Springer Verlag (2002).

[110] J. Kerr. *On the magnetism of light and the illumination of magnetic lines of force.* Rep. Brit. Ass., **S5**, 85 (1876).

[111] J. Kerr. *On the rotation of the plane of polarization by reflection from the pole of a magnet.* Phil. Mag., **3**, 321 (1877).

[112] P. N. Argyres. *Theory of the Faraday and Kerr effects in ferromagnetics.* Phys. Rev., **97**, 334 (1955).

[113] K. H. Hellwege, O. Madelung (ed.). *Landolt-Börnstein - Magnetic Properties of Metals, New Series III/19 a.* Springer Verlag, Berlin (1986).

[114] H. Niedrig (ed.). *Bergmann - Schäfer, Lehrbuch der Experimentalphysik - Optik.* Walter de Gruyter, Berlin (1993).

[115] M. R. Freeman, W. K. Hiebert. *Stroboscopic Microscopy of Magnetic Domains* In: *Spin Dynamics in Confined Magnetic Structures I,* edited by B. Hillebrands and K. Ounadjela. Springer Verlag (2002).

[116] M. Buess, Y. Acremann, A. Kashuba, C. H. Back, D. Pescia. *Pulsed precessional motion on the 'back of an envelope'.* J. Phys.: Condens. Matter, **15**, R1093 (2003).

[117] M. Buess, R. Höllinger, T. Haug, K. Perzlmaier, U. Krey, D. Pescia, M. R. Scheinfein, D. Weiss, C. H. Back. *Fourier transform imaging of spin vortex eigenmodes.* Phys. Rev. Lett., **93**, 077207 (2004).

[118] Y. Acremann, M. Buess, C. H. Back, M. Dumm, G. Bayreuther, D. Pescia. *Ultrafast generation of magnetic felds in a Schottky diode.* Nature, **414**, 51 (2001).

[119] A. V. Kimel, A. Kirilyuk, P. A. Usachev, R. V. Pisarev, A. M. Balbashov, T. Rasing. *Ultrafast non-thermal control of magnetization by instantaneous photomagnetic pulses.* Nature, **435**, 655 (2005).

[120] M. van Kampen, C. Jozsa, J. T. Kohlhepp, P. LeClair, L. Lagae, W. J. M. de Jonge, B. Koopmans. *All-optical probe of coherent spin waves.* Phys. Rev. Lett., **88**, 227201 (2002).

[121] D. Meschede. *Optik, Licht und Laser.* B. G. Teubner, Stuttgart (1999).

[122] M. Djordjevic, M. Lüttich, P. Moschkau, P. Guderian, T. Kampfrath, R. G. Ulbrich, M. Münzenberg, W. Felsch, J. S. Moodera. *Comprehensive view on ultrafast dynamics of ferromagnetic films.* Phys. Stat. Sol. C, **3**, 1347 (2006).

[123] *RegA amplifier manual.* Coherent Inc.

[124] W. K. Hiebert, A. Stankiewicz, M. R. Freeman. *Direct observation of magnetic relaxation in a small permalloy disk by time-resolved scanning kerr microscopy.* Phys. Rev. Lett., **79**, 1134 (1997).

[125] C. E. Patton, Z. Frait, C. H. Wilts. *Frequency dependence of the parallel and perpendicular ferromagnetic resonance linewidth in permalloy films, 2-36 Ghz.* J. Appl. Phys., **46**, 5002 (1975).

[126] S. Choi, K. Lee, S. Kima. *Spin-wave interference.* Appl. Phys. Lett., **89**, 062501 (2006).

[127] G. N. Kakazei, P. E. Wigen, K. Y. Guslienko, V. Novosad, A. N. Slavin, V. O. Golub, N. A. Lesnik, Y. Otani. *Spin-wave spectra of perpendicularly magnetized circular submicron dot arrays.* Appl. Phys. Lett., **85**, 443 (2004).

[128] M. Djordjević. *Magnetization dynamics in all-optical pump-probe experiments: spin wave modes and spin-current damping.* PhD thesis, University of Göttingen, Germany (2006).

[129] G. Counil, J.-V. Kim, T. Devolder, C. Chappert, K. Shigeto, Y. Otani. *Spin wave contributions to the high-frequency magnetic response of thin films obtained with inductive methods.* J. Appl. Phys., **95**, 5646 (2004).

[130] M. Kießling. *Magnetization dynamics in rare-earth doped permalloy films.* Master's thesis, University of Regensburg, Regensburg (2006).

[131] E. P. Wohlfahrt (ed.). *Ferromagnetic Materials, Vol. 1.* North Holland Publishing Group, Amsterdam (1980).

[132] J. Stöhr, H. A. Padmore, S. Anders, T. Stammler, M. R. Scheinfein. *Principles of X-ray magnetic dichroism spectromicroscopy.* Surf. Rev. Lett., **5**, 1297 (1998).

[133] E. Bauer. *Photoelectron microscopy.* J. Phys.: Condens. Matter, **13**, 11391 (2001).

[134] B. T. Tonner, G. R. Harp. *Photoelectron microscopy with synchrotron radiation.* Rev. Sci. Instrument., **59**, 853 (1988).

[135] C. Quitmann, U. Flechsig, L. Patthey, T. Schmidt, G. Ingold, M. Howells, M. Janousch, R. Abela. *A beamline for time resolved photoelectron microscopy on magnetic materials at the swiss light source.* Surf. Sci., **480**, 173 (2001).

[136] A. Scholl, H. Ohldag, F. Nolting, J. Stöhr, H. A. Padmore. *X-ray photoemission electron microscopy, a tool for the investigation of complex magnetic structures (invited).* Rev. Sci. Instrument., **73**, 1362 (2002).

[137] B. T. Thole, G. van der Laan, G. A. Sawatzky. *Strong magnetic dichroism predicted in the $M_{4,5}$ X-Ray absorption spectra of magnetic rare-earth materials.* Phys. Rev. Lett., **55**, 2086 (1985).

[138] S. S. Jaswal, D. J. Sellmyer, M. Engelhardt, Z. Zhao, A. J. Arko, K. Xie. *Electronic structure, photoemission, and magnetism in Gd_2Co and Er_2Co glasses.* Phys. Rev. B, **35**, 996 (1987).

[139] A. Kowalczyk, G. Chelkowska, A. Szajek. *X-ray photoemission spectra and electronic structure of $GdCo_4B$.* Solid State Comm., **120**, 407 (2001).

[140] J. Pelzl, R. Meckenstock, D. Spodding, F. Schreiber, J. Pflaum, Z. Frait. *Spin-orbit-coupling effects on g-value and damping factor of the ferrimagnetic resonance in Co and Fe films.* J. Phys.: Condens. Matter, **15**, 451 (2003).

[141] D. Stanciu, A. V. Kimel, F. Hansteen, A. Tsukamoto, A. Itoh, A. Kirilyuk, T. Rasing. *Ultrafast spin dynamics across compensation points in ferrimagnetic GdFeCo: The role of angular momentum compensation.* Phys. Rev. B, **73**, 220402(R) (2006).

[142] E. Stavrou, H. Rohrmann, K. Röll. *Gd/Fe multilayers with an anisotropy changing from in-plane to perpendicular for MSR applications.* IEEE Trans. Magn., **34**, 1988 (1998).

[143] E. Stavrou, K. Röll. *Magnetic anisotropy in Gd/(FeCo) and Gd/Fe multilayers for high density magneto-optical recording.* J. Appl. Phys., **85**, 5971 (1999).

[144] T. Morishita, Y. Togami, K. Tsushima. *Magnetism and structure of compositionally modulated Fe-Gd thin films.* J. Phys. Soc. Jpn., **54**, 37 (1985).

[145] T. Eimüller, E. Amaladass, B. Ludescher. *Microscopic insight into the domain configuration during a spin reorientation transition in an Fe/Gd multilayer.* to be published.

[146] W. Bailey, P. Kabos, F. Mancoff, S. Russek. *Control of magnetization dynamics in $Ni_{81}Fe_{19}$ thin films through the use of rare-earth dopants.* IEEE Trans. Magn., **37**, 1749 (2001).

Danksagung

Viele haben dazu beigetragen, diese Arbeit überhaupt erst zu ermöglichen. Dafür ein herzliches Dankeschön an ...

... Prof. Dr. Christian Back für die Möglichkeit, in seiner Arbeitsgruppe zu arbeiten, für die stetige Motivation ("I hob da g'sagt, dass UHV ned leicht is..."), die Unterstützung und das stetige Interesse am Fortgang der Arbeit.

... Dr. Georg Woltersdorf ohne den manche Messung nicht zustande gekommen wäre. Danke auch für die vielen lehrreichen Diskussionen und Erklärungen. Die sinnlose Nacht im Labor wird mir unvergessen bleiben ...

... Prof. Mike Scheinfein für sein Interesse, sein Hilfe und die Diskussionen auch über die Dinge der Physik hinaus.

... Dr. Johann Vancea für die Tiefenprofilanalysen, und Friedl Dorfner für die Röflu-Untersuchungen und diverse Aktionen im Labor (ich sag nur Vorpumpe).

... Matthias Buess, der mich in das ganze Equipment eingeführt hat und beim Aufbau des Labors unersetzlich war, und Ingo Neudecker für alles, was mit VNA-FMR zu tun hatte.

... Alexander Weber und Matthias Kießling, die beide durch ihre Diplomarbeiten wesentlich zum Gelingen der Arbeit beigetragen haben.

... den Rest der Arbeitsgruppe Back für die super Zusammenarbeit, den Spaß und die angenehme Stimmung im Labor.

... die Zimmerkollegen, die mich während meiner Zeit am Lehrstuhl begleitet haben, als da sind Wolfgang Brunner, Martin Heumann, Thomas Uhlig, Marcello Soda, Martin Beer, Martin Utz, Eva Hußnätter, Tom Haug, Karl Engl, Stephan Otto und Martin Brunner. Danke für die entspannte Atmosphäre.

... Christian Back, Georg Woltersdorf, Matthias Kiessling, Alexander Weber, Tom Haug und Tina Binder (geb. Seder) für's Korrigieren der Arbeit.

... Olga Ganicheva und Martin Beer für die Querpräparate, und Matthias Sperl für die SQUID-Messungen.

... Dieter Schierl und Tobi Stöckl für die kleinen und großen Dinge.

... die E-Werkstatt Physik für die meist schnelle und unkomplizierte Hilfe, und natürlich auch an die Kollegen aus der mechanischen Werkstatt, ohne die es auch nicht gegangen wäre.

... den Lustrat, der mich zum Lustwart gewählt hat, und so in den kleinen, aber feinen Kreis der Lusträte aufgenommen hat. Danke an Roland Meier, der als Kaffeewart mit mir unser Amtsjahr meisterte.

... alle Mitglieder des Lehrstuhls, die durch die super Zusammenarbeit, aber auch durch so manche Nachfeier, die Ausflüge und was sonst noch geboten ist, unseren Lehrstuhl zu etwas besonderem machen.

... Jörg Raabe und Christoph Quitmann für die Messungen am SLS.

... Dr. Jan Thiele, Prof. Jeff Dahn und Dr. Tim Hatchard für die Herstellung und Charakterisierung so mancher Probe.

... Oleksandr Mosendz für die BLS-Messungen und Dr. Manuel Izquierdo für die Hilfe bei CoGd.

... die DFG, die mit ihrem Sponsoring die Arbeit finanziell überhaupt erst ermöglichte.

... alle Freunde, die mich innerhalb und außerhalb der Uni unterstützt haben, und für Abwechslung sorgten.

... Tom Haug, für die unvergessliche Zeit in unserer WG, dem Studium und der Promotion.

Und zum Schluss ein großes Dankeschön an meine ganze Familie, die mir immer den nötigen Rückhalt gab und mir den Rücken frei hielt, und besonders an Tina, für ihre Liebe und Geduld ... jetzt ist's vorbei.